Gas Chromatography - Derivatization, Sample Preparation, Application

Edited by Peter Kusch

Published in London, United Kingdom

IntechOpen

Supporting open minds since 2005

Gas Chromatography - Derivatization, Sample Preparation, Application
http://dx.doi.org/10.5772/intechopen.73339
Edited by Peter Kusch

Contributors
Kristian Pastor, Marijana Ačanski, Djura Vujić, Dinesh Bilehal, Clarissa Piccinin Frizzo, Bruno Hennemann, Keli Wust, Bruna Kuhn, Nilo Zanatta, Helio Bonacorso, Marcos Martins, Alisson Paz, Foujan Falaki, Peter Kusch, Serban Moldoveanu, Victor David

First published in London, United Kingdom, 2019 by IntechOpen
IntechOpen is the global imprint of INTECHOPEN LIMITED, registered in England and Wales, registration number: 11086078, The Shard, 25th floor, 32 London Bridge Street
London, SE19SG - United Kingdom
Printed in Croatia

British Library Cataloguing-in-Publication Data
A catalogue record for this book is available from the British Library

Additional hard and PDF copies can be obtained from orders@intechopen.com

Gas Chromatography - Derivatization, Sample Preparation, Application
Edited by Peter Kusch
p. cm.
Print ISBN 978-1-83881-865-4
Online ISBN 978-1-83881-866-1
eBook (PDF) ISBN 978-1-83881-867-8

For EU product safety concerns:
IN TECH d.o.o., Prolaz Marije Krucifikse Kozulić 3, 51000 Rijeka, Croatia, info@intechopen.com or visit our website at intechopen.com.

Meet the editor

Dr. Peter Kusch studied chemistry at the Pedagogical University in Opole and received his doctorate in organic chemical technology from the Poznań University of Technology, Poland. From 1977 to 1988, he worked as an analytical chemist and adjunct at the Institute of Heavy Organic Synthesis "Blachownia" (Kędzierzyn-Koźle, Poland). After moving to Germany, he worked for several years in the *Fischer* Labor- und Verfahrenstechnik GmbH company (Meckenheim/Bonn, Germany) as laboratory manager and specialist for analytical pyrolysis and gas chromatography. Since 1998 until his retirement in 2018, he was a scientific co-worker at the Department of Applied Natural Sciences of the Bonn-Rhein-Sieg University of Applied Sciences in Rheinbach, Germany. He has been author/coauthor of over 90 scientific publications, one book, edited two books, 13 invited book chapters, and 11 patents in the area of chromatography, mass spectrometry, analytical pyrolysis, and chemical technology. Dr. Kusch is a reviewer for several international journals in the area of analytical chemistry. He is an Editorial Board Member of the journals *Polymer Testing* and *Heliyon* (both from Elsevier).

Contents

Preface

Gas chromatography (GC) is one of the most important types of chromatography used in analytical chemistry for separating and analyzing chemical organic compounds. According to the International Union of Pure and Applied Chemistry (IUPAC) recommendation, gas chromatography is defined as a separation technique in which the mobile phase is a gas. GC is a separation and detection method for sample mixtures, whose components can be volatilized without thermal decomposition. The analytical procedure is used to determine organic substances, usually molecules with a molecular mass less than 500 g/mol and a boiling point less than 400 °C. At the beginning, this method was used to study petrochemical products. Today, gas chromatography is one of the most widespread investigation methods of instrumental analysis. This technique is used in the laboratories of chemical, petrochemical, and pharmaceutical industries, in research institutes, and also in the clinical, environmental, and food and beverage analysis. Recent developments in GC have resulted in the introduction of new sample preparation techniques and developing newer separation and detection solutions, such as fast GC, multidimensional GC, and hyphenated GC.

This book is the outcome of contributions by experts in the field of gas chromatography. In Chapter 1, the definition, short history, and recent developments of gas chromatography are described.

Chapter 2 contains a detailed overview of derivatization methods in GC and GC/MS. In Chapter 3, a broad overview of sample preparation techniques in GC and GC/MS is presented. In Chapter 4, the authors present a comprehensive study on pyrazole fragmentation by gas chromatography coupled with mass spectrometry. In Chapter 5, a GC/MS/MS method has been developed and validated for the determination and quantification of 35 multiclass pesticide residues in grape samples. The QuEChERS-dSPE (Dispersive Solid Phase Extraction) method was used for the extraction of the pesticide residues. Chapter 6 contains an overview of gas chromatographic methods for food authentication.

The Editor would like to thank all the authors of these chapters for their contribution of work and commitment, which allowed the publication of this book. All the help and advice from Mrs. Marina Dusevic, the editorial Author Service Manager is also gratefully acknowledged.

Dr. Peter Kusch
Department of Applied Natural Sciences,
Bonn-Rhein-Sieg University of Applied Sciences,
Rheinbach, Germany

Section 1

Introduction

Chapter 1

Introductory Chapter: Gas Chromatography - The Most Versatile Analytical Technique

Peter Kusch

1. Definition and short history of gas chromatography

Gas chromatography (GC) is a type of chromatography. According to the International Union of Pure and Applied Chemistry (IUPAC) recommendation, gas chromatography is defined as a separation technique in which the mobile phase is a gas. Gas chromatography is always carried out in a column [1]. GC is a separation and detection method for sample mixtures, whose components can be volatilized without thermal decomposition. The analytical procedure is used for the determination of organic substances; usually molecules have a molecular mass of less than 500 g/mol and a boiling point of less than 400°C. GC is a technique used to separate mixtures of gaseous chemical compounds based on differences in the compounds' relative affinities for a solid (gas-solid chromatography) or liquid (gas-liquid chromatography) stationary phase held within a column.

Gas-liquid partition chromatography was invented by Martin and James from the National Institute for Medical Research, London, in 1952. The invention of this technique is generally attributed to the inventors in their 1952 published paper in the *Biochemical Journal* [2]. In this publication the theory of the partition column has been extended to cover a compressible mobile phase, and gas-liquid partition columns were described for the separation of volatile fatty acids. In the same year, the *Nobel Prize* in chemistry was awarded jointly to Martin and Synge for their *invention of partition chromatography*. Starting from this time, gas chromatography has become one of the most important and widely applied analytical techniques in modern chemistry. The first commercial gas chromatograph was introduced in 1955 by Perkin-Elmer (USA). Subsequently, this method was used to study petrochemical products. Today, gas chromatography is one of the most widespread investigation methods of instrumental analysis. This technique is used in the laboratories of the chemical, petrochemical, and pharmaceutical industries, in the research institutes and also in the clinical and environmental and food and beverage analyses. Recent developments in GC have resulted in the introduction of better and selective fused silica capillary columns and methods for sample preparation. Newer separation and detection solutions, such as fast GC, multidimensional separation GC techniques (GC × GC), and hyphenation of GC and GC × GC with mass spectrometry (MS), with triple quadrupole mass spectrometry and with time-of-flight mass spectrometry (TOF-MS), have been developed and become industrial routine. Analytical pyrolysis (Py) technique hyphenated to GC and GC/MS has extended the range of possible tools for the characterization of synthetic polymers and copolymers. This technique has been used extensively over the last 30 years as a complementary analytical tool used to characterize the structure of synthetic organic polymers and copolymers, polymer blends, biopolymers, and natural resins [3].

2. Multidimensional gas chromatography

Developed by Phillips and coworker at the Southern Illinois University (USA) in the early 1990s, comprehensive multidimensional GC (GC × GC or 2D GC) is a powerful technique for samples containing very large numbers of compounds of interest and also for samples which exhibit high chemical complexity. This technique can be used to separate very complex mixtures, such as those found in the petrochemical, environmental, and food and fragrance industries [4–6]. The method uses two capillary columns, typically of very different polarities, installed in series with a modulator in between. The first column is in principle nonpolar or low polar, and the second column is polar. The length of the first column might typically be 20–30 m, the inner diameter 0.25 mm, and the film thickness 0.25 μm. The second column is typically shorter (1–2 m), the inner diameter is narrower (0.1 mm), and the stationary phase is thinner (0.1 μm), to allow for faster separations. The entire assembly is located inside the GC oven [6]. The modulator collects effluent from the first column for a fraction of the time equal to peak width. The modulator focuses the material collected from each cut into a very narrow band through flow compression. It introduces the bands sequentially onto the second column, resulting in additional separation for each band injected onto the second column [4–9]. The most common data transformation is the construction of a 2D representation, in which one axis represents the separation on the first column (first dimension), and the other axis represents the secondary column separation (second dimension). Therefore, the look of GC × GC chromatograms appears completely different from conventional GC chromatogram showing a two-dimensional plane where analyte spots are scattered about [7, 8]. A contour plot, using elevation lines or color coding, represents the signal intensity. 2D GC data are primarily used for

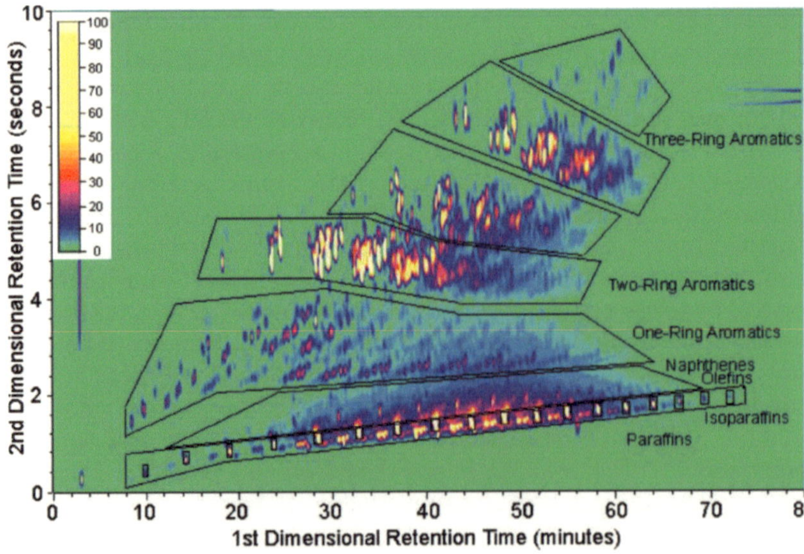

Figure 1.
2D GC plot of a refinery stream boiling at diesel temperature range. The scale indicates the relative signal intensity. Figure reprinted from Ref. [10] with permission from ACS.

qualitative analysis; however, quantitative multidimensional GC analysis is also possible [9]. **Figure 1** shows an exemplary 2D GC plot of a refinery stream boiling at diesel temperature range [10].

In this book, state of the art of gas chromatography and new developments and applications are presented. New sample preparation techniques, derivatization methods, and hyphenation with mass spectrometry are described.

Author details

Peter Kusch
Department of Applied Natural Sciences, Bonn-Rhein-Sieg University of Applied Sciences, Rheinbach, Germany

*Address all correspondence to: ptrkusch@arcor.de

References

[1] IUPAC Compendium of Chemical Terminology, International Union of Pure and Applied Chemistry, Version 2.3.3. 2014

[2] James AT, Martin AJP. Gas-liquid partition chromatography: The separation and micro-estimation of volatile fatty acids from formic acid to dodecanoic acid. The Biochemical Journal. 1952;**50**(5):679-690

[3] Kusch P. Pyrolysis—Gas Chromatography/Mass Spectrometry of Polymeric Materials. London, UK: World Scientific Europe; 2018

[4] Liu Z, Phillips JB. Comprehensive two-dimensional gas chromatography using an on column thermal modulator interface. Journal of Chromatographic Science. 1991;**29**(6):227-231

[5] Phillips JB, Xu J. Comprehensive multi-dimensional gas chromatography. Journal of Chromatography A. 1995;**703**:327-334

[6] Comprehensive GC System Based on Flow Modulation for the 7890A GC, Application Brief. Wilmington, DE, USA: Agilent Technologies; 2008

[7] Hinshaw JV. Comprehensive two-dimensional gas chromatography. LCGC Europe. 2004;**17**(2):86-95

[8] Mondello L. GC × GC Handbook, Fundamental Principles of Comprehensive 2D GC. Japan: Shimadzu; 2012

[9] Taylor T. A short introduction to multidimensional GC. LCGC North America. 2012;**30**(9):870

[10] Wang FC, Qiang K, Green LA. GCxMS of diesel: A two-dimensional separation approach. Analytical Chemistry. 2005;**77**(9):2777-2785

Section 2

Derivatization Methods

Chapter 2

Derivatization Methods in GC and GC/MS

Serban C. Moldoveanu and Victor David

Abstract

The first part of this chapter presents the main objectives for performing derivatization of a sample to be analyzed by gas chromatography (GC) or gas chromatography/mass spectrometry (GC/MS). The derivatization is typically done to change the analyte properties for a better separation and also for enhancing the method sensitivity. In GC/MS, derivatization may improve the capability of compound identification. Examples illustrating such improvements are included. The second part describes several types of derivatization that are more frequently used in analytical practice. These include alkylation (e.g., methylation), formation of aryl derivatives, silylation (e.g., formation of trimethylsilyl derivatives), acylation (e.g., reactions with acyl chlorides or with chloroformates), and several other types of derivatizations. The chapter also presents typical derivatizations for analytes with specific functional groups and discusses artifact formation in certain derivatization reactions.

Keywords: gas chromatography, mass spectrometry, derivatization, alkylation, aryl derivatives, silylation, acylation

1. General comments

Two specific trends can be noticed in modern chemical analysis. One is the continuous demand for more sensitive and accurate analytical methods. The other is the desire for simpler methods that require as little as possible human intervention. One of the various procedures to make the analytical methods more sensitive and accurate is the use of specific chemical changes (e.g., derivatization) applied on the analytes or even on the whole sample. However, these changes frequently involve more human intervention than the direct use of advanced instrumentation. For this reason, the methods involving chemical changes such as derivatizations are not necessarily the first choice when selecting an analytical method. Nevertheless, in many cases, the benefits of derivatization are more important than the disadvantage of requiring human intervention, and for this reason, derivatization is still frequently used in the analytical practice. Also, modern GC, GC/MS (or GC/MS/MS) instrumentation may offer autosampling with the capability of adding reagents to the sample, as well as stirring, heating, and injecting the sample at specific time intervals in the GC system. This type of instrumentation may reduce significantly the human handling involved in derivatization.

Various chemical changes can be performed on an analyte in order to make it suitable for a specific method of analysis. The most common is derivatization, but other chemical changes can be utilized, for example, pyrolytic decomposition and, in the case of polymers, polymer fragmentation using reagents. The choice depends on the nature of the analyte, the sample matrix, the intended changes in the analyte properties, and the analytical method to be used.

The addition of a reagent on a sample may produce a chemical reaction only with the analytes without affecting the matrix. However, it is also possible that some matrix components are derivatized unintentionally. Usually, it is preferable to have only the analytes derivatized since in this way a better separation from the matrix is expected. Some derivatizations are used in the sample cleanup or concentration process. Also, the derivatization process may be combined with simultaneous extraction and concentration of the sample or may be followed by a second preparation step before the chromatographic analysis. More frequently, the derivatization is done to change the analyte properties for the core analytical procedure (GC, GC/MS, etc.).

Derivatization can be applied before the core chromatographic process or after it. Precolumn derivatization takes place before the separation and postcolumn derivatization after it. In GC precolumn derivatization is much more common and most derivatizations are performed "offline." There are however derivatizations that can be done "online," for example, in the injection port of the GC such as some methylations using tetramethyl ammonium hydroxide (TMAH). Postcolumn derivatizations are performed only for enhancing the detectability of the analytes. Typically, they must be done "online" and should be completed in the specific time frame needed by the analyte to reach the detector.

A wide variety of derivatization reagents and procedures are described in the literature, with the reagents carrying specific moieties that provide a desired property to the analytes, as well as with specific reactive groups that permit the reaction with the analyte. Multiple step derivatizations as well as derivatizations followed by a second one are known.

Derivatization is not always the first step in sample preparation. Sample preparation typically includes other operations, besides derivatization. Some of these steps are more complex such as sample cleanup or concentration and others more simple such as pH adjustments, addition of proton acceptors or donors, change of the medium (from one solvent to another), and addition of catalysts to enhance the derivatization, and these may be necessary for a successful derivatization.

Although derivatization is performed in order to make possible or to improve the results of a chemical analysis, there are also some disadvantages of using derivatization. Besides the potential need of more manpower for the analysis, the addition of more operations applied on the sample (including the analytes) can be a source of additional errors. In particular the involvement of a chemical reaction that may not be perfectly controlled can bring significant errors in the analytical results. To minimize the potential errors when using derivatization, specific aspects of the derivatization must be considered in its choice, such as the efficiency of the chemical reaction used in the derivatization, the stability of the derivatized analytes, the availability of reagents and necessary equipment, and the time necessary for performing the analysis. For a given analyte or group of analytes, the reaction with the derivatization reagent must be complete or at least close to complete, must take place in a length of time that is not prohibitive, and must have very little loss of the analyte with formation of artifacts or decomposition products. Only when such criteria are satisfied can a specific chosen derivatization be applied successfully.

The application of derivatization in chromatography is the subject of many studies. Numerous derivatizations have been reported in journals (e.g., *J. Chromatogr. A* and *B*, *J. Chromatogr. Sci.*, *J. Sep. Sci.*, *Chromatographia*, etc.), in various books [1–5], in application notes of instrument manufacturers, as well as on the web.

2. Derivatization for improving separation in gas chromatography

For GC analysis, the effect of derivatization can be beneficial in a variety of circumstances. Some of the most common uses of derivatization for improving the GC separation are the following:

(a) Derivatization that replaces active (polar) hydrogen atoms in the analyte to decrease its boiling point. The active hydrogens in a chemical compound typically enhance the capability to form hydrogen bonds and increase the compound polarity. For this reason, many compounds containing active (polar) hydrogens are not volatile, the volatility being necessary for using GC or GC/MS as a core analytical method. Derivatization can be used to replace active hydrogens from an analyte Y-H (or Y:H) in functional groups such as OH, COOH, SH, NH, and CONH. These reactions can be written in a simplified form as follows:

$$\text{Y-H} + \text{R} - \text{X} \rightarrow \text{Y-R} + \text{HX} \tag{1}$$

In reaction (1), the reagent R-X contains an "active" group X and a group R that carries a desired property (e.g., lack of polarity for GC). Group R in the reagent can be a low molecular mass fragment such as CH_3 or C_2H_5, a short-chain fluorinated alkyl in alkylation reactions, $Si(CH_3)_3$ or other silyl groups in silylations, $COCH_3$ or short-chain fluorinated acyl groups in acylations, etc. An example of a chromatogram resulting from the GC/MS analysis of a silylated tobacco sample is given in **Figure 1**. Tobacco contains many hydroxy acids such as malic, trihydroxybutanoic, citric, quinic, glucuronic, and chlorogenic. Also, it contains monosaccharides (e.g., glucose, fructose), disaccharides (e.g., sucrose), and even trisaccharides. None of these compounds are volatile, having numerous active hydrogens. The replacement of these hydrogens with $Si(CH_3)_3$ by silylation renders these compounds volatile, and they can be analyzed by GC/MS as seen in **Figure 1**.

(b) Derivatization for enhancing the separation. Specific moieties added to an analyte may be necessary for enhancing the separation. This is frequently practiced for general GC separations and is also very useful for the separation of chiral molecules (see Section 4). The derivatized analytes may have significantly different properties from each other, for example,

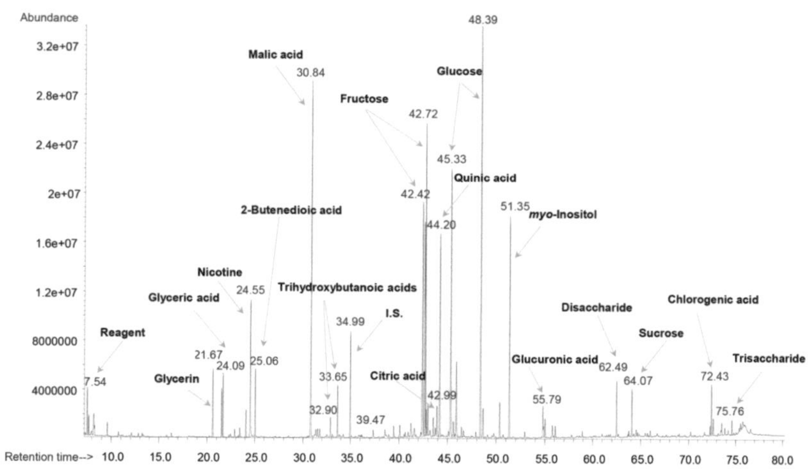

Figure 1.
GC/MS chromatogram of a silylated tobacco sample, with separation on a DB-5 MS column from Agilent (Agilent Technologies Inc., Wilmington, DE, USA) (Note: an internal standard I.S. was added to the sample).

regarding polarity and implicitly in their boiling point, allowing separations that are difficult to achieve otherwise. Also, derivatization may generate more significant differences between the analytes and the matrix components.

(c) Derivatization that replaces active hydrogens in the analyte to improve the behavior of the analyte in the chromatographic separation. The chromatographic column (e.g., a capillary column coated with a bonded stationary phase) may display additional capability to interact with polar molecules, besides the intended interactions due to its bonded phase. This may come, for example, from the silica wall of the column. Secondary interactions taking place with only a portion of the molecules of the analyte generate peak tailing. This is exemplified in **Figure 2** which shows a hypothetical case of two different types of interaction between the column and a specific molecular species.

(d) Derivatization for the improvement of stability of a compound. This stability may refer to thermal stability, a property which overlaps to a certain extent to what was described at point (a). However, even some volatile compounds may be further thermally stabilized by derivatization. Also, chemical stability can be enhanced by protecting specific groups in the analyte using derivatization. For example, thiols can be protected using derivatization against oxidation by the traces of oxygen in the heated injection port of the GC.

The choice of the appropriate derivatization is not always a simple task. The replacement of a hydrogen atom with a group of atoms may increase the molecular weight of the derivatized analyte. In such cases, it must be verified that the increase in the molecular weight by derivatization brings no or only a small increase in the boiling point of the analyte. Most of the time, low molecular weight substituents such as CH_3 or $Si(CH_3)_3$ are preferable for GC analysis to the active hydrogens for achieving the previously described goals. Large substituents may increase the boiling point too much and make the compound not acceptable for GC analysis.

Besides replacement of active hydrogens, other derivatization reactions can be utilized. For example, condensation reactions may decrease the boiling point and improve the thermal stability of an analyte. However, the generation of new active hydrogens must be avoided in condensation reactions or must be followed by a second derivatization.

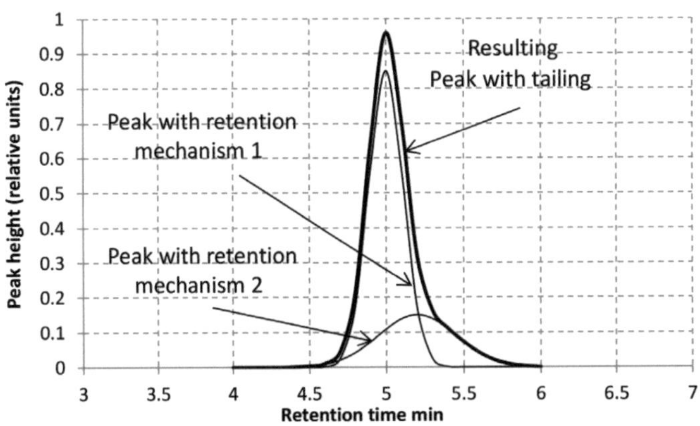

Figure 2.
Peak tailing due to multiple retention mechanisms.

3. Derivatization for chiral separation in gas chromatography

The compounds with structures that are mirror images to each other are indicated as enantiomers, and their molecules are not superimposable, having the property called chirality. Chirality is commonly caused by the existence in the molecule of at least one tetrahedral carbon atom substituted with groups that are different. However, chiral molecules may be generated with a phosphorus or a sulfur chiral atom. Not only chiral centers (such as an asymmetric carbon) generate enantiomers, but a chiral axis or a chiral plane can lead to enantiomers. The chirality in an enantiomer is specified using the symbols R and S based on specific rules. For the assignment of a symbol R or S to a chiral carbon, the substituents are arranged in a sequence $a > b > c > d$. For the four atoms directly attached to the asymmetric carbon, a higher atomic number outranks the lower, and a higher atomic mass outranks the lower mass. For the same atoms directly attached to the asymmetric carbon, the priorities are assigned at the first point of difference. After the sequence is established, the molecule is oriented in space with the group "d" of the lowest priority behind the asymmetric carbon. When viewed along the C—d bond (from C) and the three substituents a, b, and c are oriented clockwise, the compound contains an R asymmetric carbon, and it contains an S asymmetric carbon for counterclockwise arrangement.

More than one asymmetric carbon can be present in a molecule, as in the case of carbohydrates. The stereoisomers generated by more than one asymmetric carbon can be mirror image one to the other (enantiomers) or may have different steric arrangements being diastereoisomers. These types of molecules are schematically shown in **Figure 3**.

The (S,S)- and the (R,R)-compounds from **Figure 3** are enantiomers, while the (S,R)-compound is a diastereoisomer to both (S,S)- and to (R,R)-compounds (it is an enantiomer to the (R,S)-compound). The gas chromatographic separation of enantiomers can be done only using chromatographic columns having chiral stationary phases. The derivatization of enantiomers with non-chiral reagents generates molecules that remain enantiomers. This type of derivatization may improve the chromatographic separation from other molecules, but the derivatized compounds of remaining enantiomers cannot be separated except on chiral stationary phases. Sometimes, better separation can be obtained even between the enantiomers (on chiral chromatographic columns) after derivatization. One such example is the separation of (R)- and (S)-nornicotine derivatized with isobutyl chloroformate on a chiral Rt-BDEXsm column with separation improved compared to that of underivatized enantiomers [6]. The derivatization reaction is indicated below:

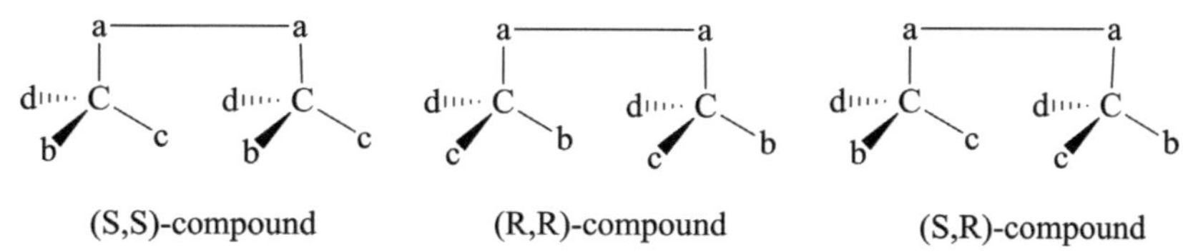

Figure 3.
Compounds with two chiral centers.

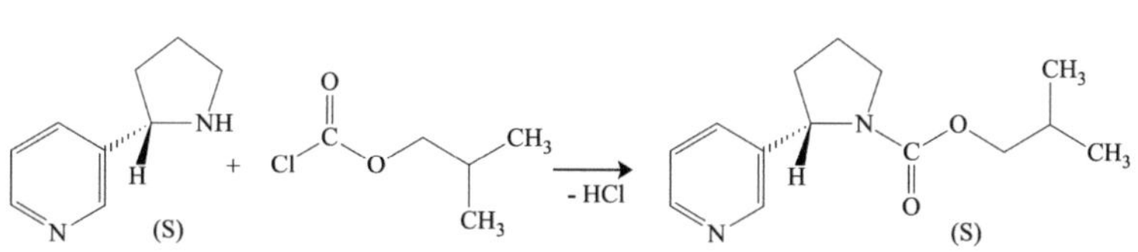

(2)

Diastereoisomers can be separated on chromatographic columns with non-chiral stationary phases which offer a much wider possibility to select the column. For this reason, an alternative procedure toward the separation of enantiomers is using derivatization with chiral reagents. This type of derivatization generates diastereoisomers which can be separated on non-chiral stationary phases.

A discussion on the separation of enantiomers on chiral phases without derivatization is beyond the purpose of this chapter. Numerous publications are dedicated to this subject, including papers published in general chromatography journals or in dedicated journals (e.g., *Chirality*), books (see, e.g., [7]), and information on the web.

The separation after derivatization with a pure enantiomer reagent is based on formation of diastereoisomers that can be separated on regular stationary phases. Depending on the nature of the analyte and of the derivatization, different separation techniques can be applied. A variety of common columns are used for such GC separations. The choice of the column is again dependent on the analyte and the derivatization procedure. For example, α-substituted organic acids such as α-chloropropionic, α-bromocaproic, etc. can be derivatized with a specific enantiomer of an amino acid ester (e.g., ethyl 2-aminopropanoate) in the presence of a peptide coupling reagent (benzotriazol-1-yl-oxy-tris(dimethylamino)-phosphonium hexafluorophosphate or BOP) in a reaction of the type:

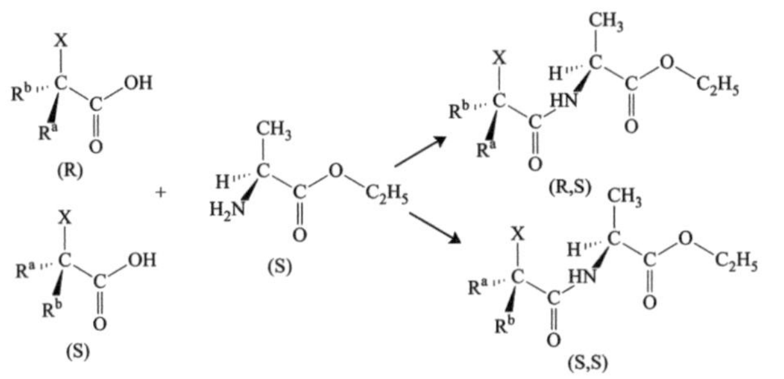

(3)

The derivatized acids that are now diastereoisomers (R,S) and (S,S) can be separated on a common capillary column (e.g., a DB-1701 column from Agilent). Another example of derivatization with a chiral reagent is that of methamphetamines with (R)-menthyl chloroformate. This derivatization allows the separation of over-the-counter (R)-methamphetamine from the illicit (S)-methamphetamine. The reaction of the (R)-enantiomer is indicated below [8]:

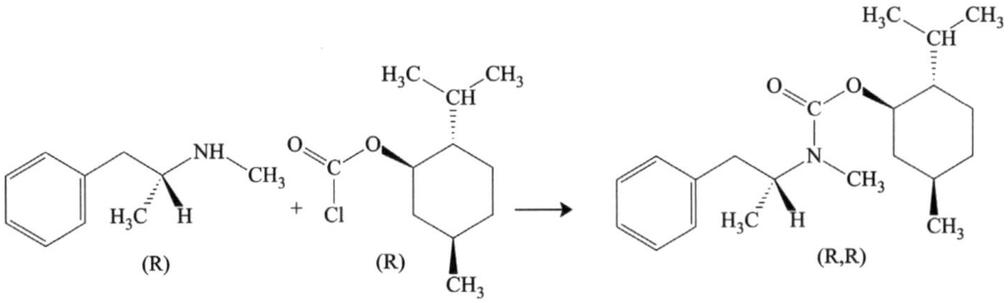

(4)

The separation of the (R,R) and (S,R) derivatives was possible on a non-chiral column for a GC/MS analysis.

4. Derivatization for improving gas chromatographic detection with other detectors than MS

Gas chromatography (not coupled with mass spectrometry, GC/MS being separately presented) used as an analytical technique can involve various detectors. The variety of such detectors is rather large, and several types include the following: thermal conductivity detector (TCD), flame ionization detector (FID), nitrogen-phosphorus detector (NPD), electron capture detector (ECD), flame photometric detector (FPD), photoionization detector (PID), electrolytic conductivity (Hall), sulfur chemiluminescence, nitrogen chemiluminescence, aroyl luminescence detector (ALD), atomic emission detector (AED), helium ionization detector (HID), vacuum ultraviolet (VUV) absorbance, infrared Doppler (IRD) absorption, FID with catalytic conversion of all analytes in CH_4 (e.g., Polyarc system [9]), etc. The derivatization with the purpose of improving detectability in GC is determined by the type of detector utilized. Most derivatizations are performed precolumn, even if they are applied only with the purpose of improving detection. However, it is important that the derivatization for improving detection does not deteriorate the separation. Preferably, both the detection and the chromatographic separation are improved by the same derivatization. Some specific postcolumn reactions applied to the analytes are part of certain types of detectors such as chemiluminescence detectors, atomic emission detectors (AED), and FID with catalytic conversion into CH_4. Some of these chemical changes in the analytes are not necessarily classified as derivatization reactions.

No specific derivatization is usually recommended to improve sensitivity when using nonselective detectors such as TCD and FID. However, in some cases when the detector is not sensitive to a specific analyte, such as formaldehyde or heavily halogenated compounds, derivatization can be used to enhance detection.

In case of NPD detector, derivatization with nitrogenous compounds can be done, which should give a higher sensitivity. However, this type of derivatization is not very common. An adverse result occurs for the NPD detectors when silylation is performed on the sample. Besides a possible reduction in the NPD response on silylated compounds containing nitrogen, a drastic decrease in the lifetime of the detector may occur, probably due to the excess of silylating reagent that commonly is injected with a derivatized sample and affects the alkali active element of the NPD.

The response of the photoionization detector (PID) depends on the ionization potential of the analyte, and compounds with higher ionization potential are not sensitive in PID, while those with lower ionization potential may have excellent sensitivity, as low as 10^{-12} mg of sample. A derivatization resulting in lowering the ionization potential of the analyte may be beneficial for PID detection. However, derivatization for enhancing PID response is not frequently used.

Some detectors such as electron capture detectors (ECD) may benefit very much from certain derivatization types. ECD (as well as negative chemical ionization mass spectrometry or NCI-MS) can be extremely sensitive, but they are selective to compounds that are able to form more stable negative ions. ECD, for example, can have sensitivity as low as 10^{-13} mg of analyte in the detector compared to the best sensitivity of FID that can be 10^{-8} to 10^{-11} mg of analyte. The efficiency of the process seems to be related to the ease of attaching an electron on the molecule. In ECD this process can be written as follows:

$$A + e^{-} \rightarrow A^{-} \tag{5}$$

With some exceptions, ECD response can be correlated with the electron affinity of the analyte [4]. In general, the halogen substituents increase the sensitivity in ECD in the order I > Br > Cl > F. Multiple substitutions seem to have a cumulative effect. Besides halogens, nitro groups seem to have an effect similar to chlorine groups. For aromatic compounds, the substituents affect the sensitivity of the ECD according to their electron withdrawing capability. Strong electron withdrawing groups such as NO_2 increase the sensitivity of the detection, while electron donating groups reduce it.

A variety of substitution groups containing electronegative elements (halogens) or nitro groups can be attached to an analyte. The procedure to attach these groups is in most cases the typical substitution of an active hydrogen in the analyte Y-H with a group R from a reagent R-X that has the appropriate active X group. Some groups used for enhancing ECD (as well as NCI-MS) sensitivity following an alkylation or aryl derivatization reaction are shown in **Figure 4**, and several substitution groups introduced by acylation, chloroformylation, or sulfonation used for the same purpose are shown in **Figure 5**. Besides alkylation or aryl derivatization, other derivatization techniques used to replace an active hydrogen are applied to introduce into a molecule as a substituent containing halogens or nitro groups enhancing

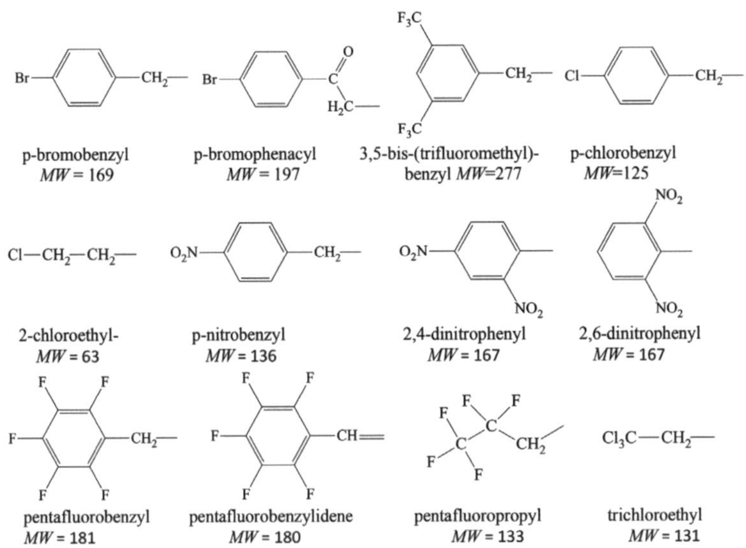

Figure 4.
Substitution groups used in alkylation and aryl derivatization for enhancing ECD (and NCI-MS) detectability (the masses are considered only for the most abundant isotope.).

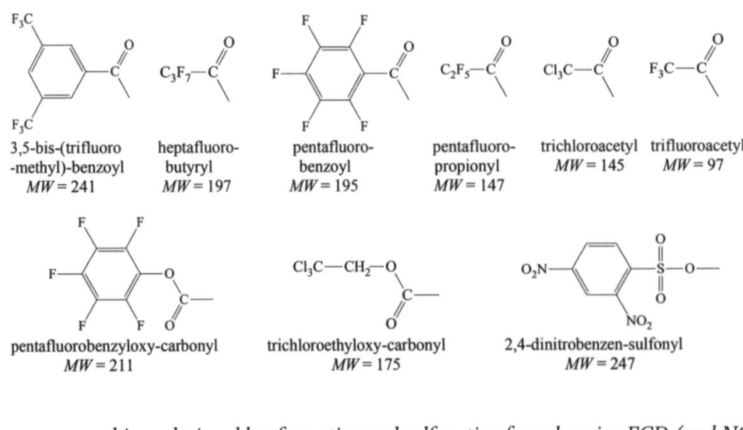

Figure 5.
Substitution groups used in acylation chloroformation and sulfonation for enhancing ECD (and NCI-MS) detectability.

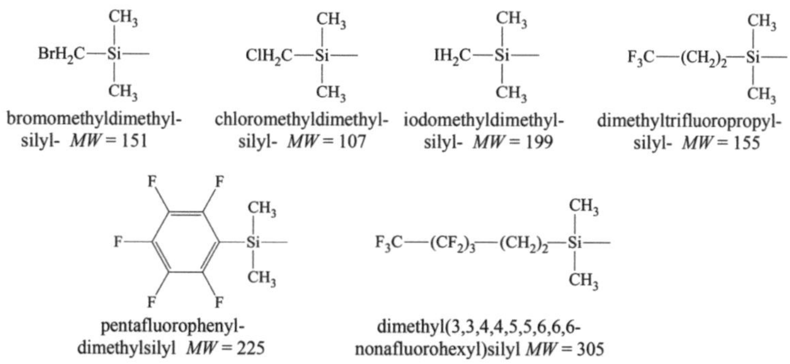

Figure 6.
Substitution groups used in silylation for enhancing ECD (and NCI-MS) detectability.

significantly the detectability of the derivatized analytes by ECD (as well as NCI-MS). Silylation, for example, can be used for this purpose when silyl groups used for derivatization contain halogens. Several silyl groups containing halogens that can be attached to an analyte by silylation with special reagents are given in **Figure 6** [4].

5. Derivatization for improving GC/MS qualitative and quantitative analysis

The most powerful tool used for compound identification purposes is very likely mass spectrometry (spectroscopy). This technique is capable to provide information from very low amounts of material such as that eluting from a chromatographic column and can be easily coupled with a gas chromatograph. Most analyses performed with MS detection (GC/MS or GC/MS/MS) are using EI+ ionization mode with electron impact at 70 eV. The electrons interact with the molecule A to eject an additional electron leaving a positively charged species (with an odd number of electrons) of the type $A^{\blacksquare+}$. The ions also receive energy during electron impact and the excess of energy induces fragmentation. For most molecules, this process can be written as follows:

$$A + e^- \rightarrow A^{\blacksquare+} + 2e^- \text{ and } A^{\blacksquare+} \rightarrow B_i^+ + C_i^{\blacksquare} \quad (6)$$

The fragments B_i^+ are commonly but not always with an even number of electrons. The formation of molecular ions takes place with a range of internal energies, and more than one fragmentation path is possible for a given molecule. Also, the fragments can suffer further fragmentations. In general, the most abundant fragment ion results from the fragmentations that form the most stable products (ion and neutral radical). The abundance of a fragment ion is affected by its stability. For this reason, the intensity of the response of a mass spectrometric detector can be very different for different molecular species, and the prediction of this intensity is difficult. As a result, the improvements in the sensitivity in EI + −type mass spectrometry (in GC/MS using EI+ ionization) are not usually sought (but not impossible) through derivatization.

Derivatization for enhancing sensitivity is, however, frequently applied in NCI-MS. In this technique, the electrons interact with the molecules of the CI gas which is lowering their energy but without forming ions. The ionization of analyte molecules takes place by interaction with the low-energy electrons or with already formed negative ions by electron capture, dissociative electron capture, ion pair formation, or ion molecule reaction. The ionization process with the formation of negative ions is efficient only for molecules with positive electron affinities. For this reason, the sensitivity in NCI-MS is highly dependent on the electron affinity of the analyte, similarly to the sensitivity in ECD. For enhancing the electron affinity, the derivatization with reagents containing, for example, fluorinated moieties (indicated in **Figures 4–6**) is practiced. The sensitivity of the analytical methods where such derivatization is applicable can have very good sensitivity. For example, derivatization with heptafluorobutyric anhydride of aromatic amines that are present at low trace level in cigarette smoke leads to limit of detection (LOD) values as low as 0.05 ng/cig. for compounds such as 4-aminobiphenyl [10, 11].

The fragmentation pattern generated by EI+ ionization mode that generates a specific mass spectrum of a molecule is very likely the most utilized technique for the identification of the molecular species. For this identification, large libraries of mass spectra are available, and computer algorithms are used for automatic searches. The identification of compounds using mass spectroscopy is not a simple process even with the capabilities offered by the electronic searches in the mass spectral libraries. This is particularly true for analysis of complex mixtures or when the analyzed compound is present in traces. Some compounds do not have a very characteristic mass spectrum, or during the chromatographic process, the separation is not achieved, and it is difficult to make an identification due to the spectra overlapping. Also, numerous compounds may have a mass spectrum that matches more than one compound (with a good quality fit). In such cases, a derivatization with the purpose of obtaining a compound that forms more informative fragments in the mass spectrum can be very useful.

The fragments from derivatized compounds can be used for the identification of unknown compounds using library searches and even when the mass spectrum is not available in the libraries. As an example, the derivatization by silylation allowed the identification of a new pentacyclic triterpenoid present in several bioactive botanicals [12]. An unidentified compound with MW = 456.7 was detected by LC/MS/MS in a rosemary extract. The structure of the compound was elucidated after silylation of the plant material based on the comparison of mass spectrum of the unidentified compound with that of silylated betulinic acid. The new compound was identified as (3β)-3-hydroxy-lupa-18,20(29)-dien-28-oic acid (or betul-18-en-oic acid). The mass spectra of the two acids are shown in **Figure 7**.

The two mass units difference between different fragments from the mass spectra of the two compounds allowed the identification of the new compound structure. Neither free betulinic acid nor betul-18-en-oic acid are volatile, such that the use of GC/MS for identification was possible only after derivatization.

Another special procedure that may be utilized for compound identification based on mass spectra is the use of two parallel derivatizations, one of them being done with an isotope-labeled reagent. Common labeling isotopes are ^{2}H (deuterium, d), ^{13}C, ^{15}N, etc. One such isotopic labeling can be done, for example, using silylation with d_{18}-N,O-bis(trimethylsilyl)-trifluoroacetamide (d_{18}-BSTFA). Derivatization of an aliquot of sample with regular BSTFA and another with d_{18}-BSTFA provides a pairing chromatogram with peaks at retention times that have only small differences from the first but with spectra differing by a number of units. The comparison of the spectra for corresponding peaks (based on retention time) of a given compound allows the calculation of the number of silyl groups attached to that compound. In addition, the fragmentation in the spectra can be better interpreted allowing easier compound identification.

Derivatization in GC/MS analysis may have multiple other utilizations and benefits. For example, quantitative analysis frequently utilizes isotopically labeled internal standards. In an analysis with multiple analytes, addition of an isotopically labeled internal standard for each analyte may become a complex process. When a

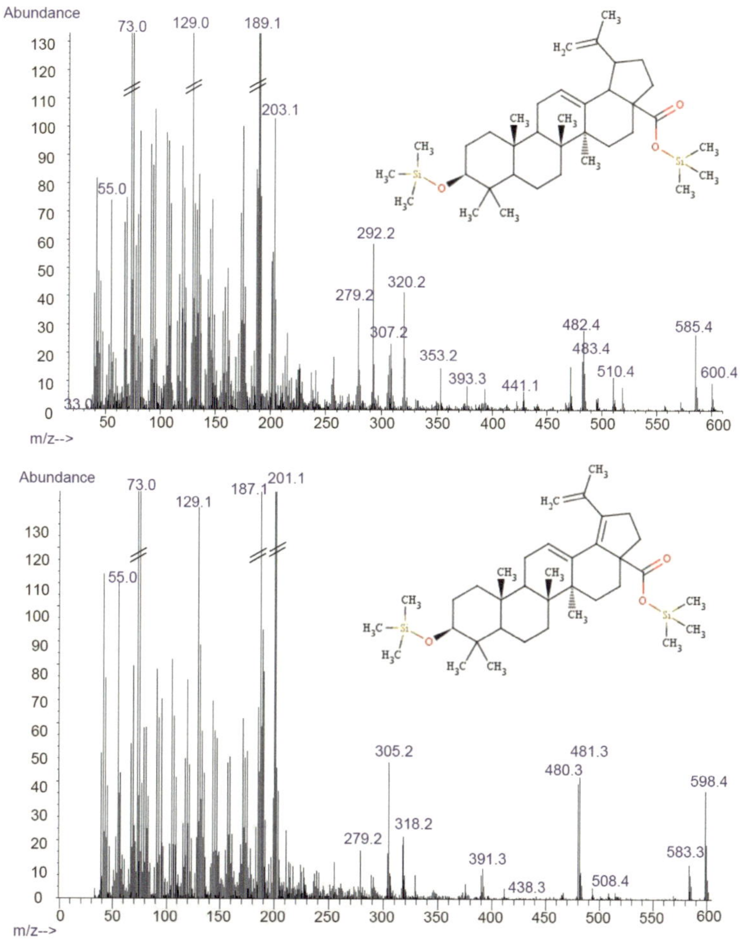

Figure 7.
Mass spectrum of silylated betulinic acid and that of silylated betul-18-en-oic acid.

derivatization is involved in the analysis, this can be done with a non-labeled reagent for the analytes in the sample, while the internal standards are obtained by derivatization of standards with the same reagent but isotopically labeled. Such technique has been proven to be very successful, for example, in the analysis of multiple amino acids (but using an LC/MS/MS procedure [13]).

6. General comments regarding the main types of chemical reactions used in derivatization

Derivatizations as chemical reactions can be classified as follows: (1) reactions with formation of alkyl or aryl derivatives, (2) silylation reactions, (3) reactions with formation of acyl derivatives, (4) reactions of addition to carbon-hetero multiple bonds, (5) reactions with formation of cyclic compounds, and (6) other reactions specific to a certain analysis. The selection of the derivatization reaction is typically done based on the desired property to be brought to the analyte and its possible reactivity. For this reason, the reagent is selected to have moieties that add the desired property to the analyte and also to have the capability to react with the specific functional group of the analyte. The matrix of the sample also has a role in

Properties	Increased acid character → ← Increased nucleophile character				
Compound	Amine	Amide	Alcohol	Phenol	Acid
First derivatization preference	Acylation	Acylation	Silylation	Silylation	Alkylation
Second derivatization preference	Alkylation	Alkylation	Acylation	Acylation	Silylation

Table 1.
Derivatization preferences for compounds containing active hydrogens.

the choice of a specific derivatization procedure. Initial matrix of the sample is not always suitable for derivatization, and in some cases preliminary sample preparation is necessary to change this matrix. The change can be as simple as drying the initial sample but can also be rather complex [14]. **Table 1** gives a simplified view of preferences for the choice of a derivatization reagent for compounds containing active hydrogens [14].

Besides functionalities with active hydrogens, other functionalities can also be derivatized. Compounds containing carbonyls can be derivatized, for example, using condensation reactions. Some analytes may contain multiple functional groups such as the amino acids. Specific derivatization reactions can be selected for such cases.

7. Reactions with formation of alkyl or aryl derivatives

The formation of alkyl or aryl derivatives is applied to replace the active hydrogens from an analyte with an alkyl (R) or aryl (Ar) group. The replacement can be done in functionalities such as OH, COOH, SH, NH, or CONH. For example, the derivatization with short-chain alkyl bromides or iodides has numerous analytical applications for compounds such as steroids, amino acids, catecholamines, sulfonamides, phenols, barbiturates, organic acids, and mono- and oligosaccharides. A large number of reagents R-X are known, and in a simplified approach, it can be considered that R is carrying a specific property and X a specific reactivity, although the reactivity of a reagent is influenced by both R and X components of the molecule. The type of moiety R and that of reactive group X are guiding the selection process of selecting a reagent for a specific derivatization.

In most alkylation reactions, the analyte acts as a nucleophile (Y:, Y:H, Y:-) reacting in a substitution (SN) with the alkylating reagent R-X, which contains a leaving group X and an alkyl group R:

$$Y{:}H + R - X \rightarrow Y - R + X{:}H \quad (7)$$

Various reagents and conditions were utilized in the derivatizations for analytical purposes. As reagents R-X for alkylations, one of the most commonly used are the alkyl halides, especially alkyl iodides and alkyl bromides. Because some of the derivatizations can be slow and inefficient depending on the analyte and on the reagent, the reaction rate becomes an important parameter for the analytical applicability. The reaction with an alkyl halide for the preparation of methyl or ethyl substituents, for example, is frequently performed either with a specific methylation reagent, in the presence of a catalyst, or in some instances using a particular solvent. The enhancement of the alkylation efficiency can be achieved using several other procedures. For example, for the analytical alkylation of carboxylic acids, specific cryptands such as crown ethers can be used to solvate the alkali metal portion of an organic acid salts, allowing the anion to be freer and increasing the rate of nucleophilic substitution. One other approach for enhancing the alkylation efficiency is the use of phase transfer alkylation. This approach is based on the formation of a compound easily extractable in an organic phase and on the displacement of the equilibrium in the direction of the formation of the desired product.

One different way of enhancing the alkylation efficiency is the use of different alkylating reagents besides short-chain alkyl bromides or iodides. One example of a halide that is particularly reactive is pentafluorobenzyl bromide. This reagent can be used for the derivatization of a variety of compounds containing active hydrogens. Another reactive halide is 2-bromoacetophenone (phenacyl bromide). This reagent is used mainly for the alkylation of compounds containing more acidic hydrogens such as carboxylic acids. Another example of methylation using a special reagent R-X is applied on carbohydrates [15]. This methylation uses methylsulfinylmethanide anion. The reagent is prepared from dry DMSO and NaH or KH in a reaction as follows:

$$(CH_3)_2SO + NaH \rightarrow CH_3 - SOCH_2^{-}Na^{+} + H_2 \qquad (8)$$

A polyol or a monosaccharide dissolved in DMSO is easily methylated with methylsulfinyl-methanide anion.

Other alkylating reagents are known (different X in R-X), also reacting in a nucleophilic substitution. For example, dimethyl sulfate can be used for alkylations. Alkylfluoromethyl-sulfonates are even more reactive than sulfates, and the reaction may take place with the active hydrogen even from alcohols or amines as follows:

$$R^a{-}OH + F_3C{-}SO_2{-}OR \longrightarrow R^a{-}OR + F_3C{-}SO_3H \qquad (9)$$

$$R^aR^bNH + F_3C{-}SO_2{-}OR \longrightarrow R^aR^bRNH^{+}\,{}^{-}O_3S{-}CF_3 \qquad (10)$$

Even tertiary amines, such as pyridine, also react with this type of reagent forming quaternary ammonium salts. The alkylation with alkylfluorosulfonates can be catalyzed as other alkylation reactions for increasing the reaction rate. A catalyst that can be used in this reaction is $Hg(CN)_2$.

Diazomethane is another common alkylating (methylating) reagent. The alkylation using diazomethane is assumed to take place as follows:

$$Y{:}H + H_2C{=}\overset{+}{N}{=}\ddot{N}{:}^{-} \longrightarrow CH_3{-}N^{+}{\equiv}N{:} + Y{:}^{-} \longrightarrow Y{-}CH_3 + N_2 \qquad (11)$$

Diazomethane is a gaseous unstable substance, which cannot be stored for long periods of time. It is usually prepared in small quantities and used immediately with or without an intermediate step of dissolution in ether. The preparation can be done from different N-nitroso-N-alkyl compounds in a reaction with a base. A common preparation uses N-nitroso-N-alkyl-p-toluenesulfonamide (Diazald). Methylation with diazomethane may require addition of a Lewis acid catalyst such as BF_3. The methylation of partly acetylated sugars and amino sugars using diazomethane and BF_3 in ether leads to the methylation of the free OH groups without the migration or substitution of the existent acyl groups.

A common alkylation of acidic analytes such as carboxylic acids, phenols, and thiols is performed using another type of alkylating reagent, namely, N,N-dimethylformamide dialkyl acetals. N,N-Dimethylformamide dimethyl acetal (Methyl-8®) is commonly used for methylations. For a compound containing a COOH group, the reaction with this reagent takes place as follows:

$$Y{-}COOH + (CH_3)_2N{-}CH(OCH_3)_2 \longrightarrow Y{-}COOCH_3 + (CH_3)_2N{-}CH{=}O + CH_3OH \qquad (12)$$

The compounds with acidic hydrogens can also be alkylated (methylated) using trimethyl orthoacetate, alkyl-p-tolyltriazenes ($R{-}NH{-}N{=}N{-}C_6H_4{-}CH_3$), and O-alkyl isoureas are also used for the formation of analytes containing acidic hydrogens, imino esters, etc.

Alcohols can also act as alkylating reagents in particular when the analyte contains a more acidic hydrogen. Catalyst such as HCl, BF_3, CF_3 COOH or a cation exchange resin in H^+ form is also frequently added to facilitate the reaction. The addition of HCl can be made as a water solution or as gaseous HCl that does not bring additional water to the reaction medium. The formation of alkyl or aryl derivatives of acids is a particularly important reaction known as esterification. Derivatization by esterification has been used with acids as the analyte and the alcohol as the reagent and also with the alcohol as the analyte and the acid the reagent. The esterification can be viewed either as the acid alkylation or as the acylation of the alcohol (see also the esterification mechanism). This reaction is typically catalyzed by strong acids and can be written as follows:

$$R{-}COOH + R^a{-}OH \xrightarrow{+H^+} R{-}COOR^a + H_2O \qquad (13)$$

The mechanism of ester formation can be summarized by the following series of reactions:

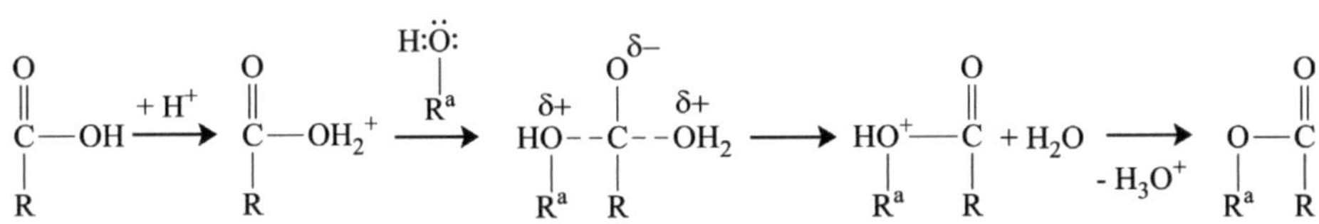

(14)

The esterification efficiency can be improved by removing the water formed in this reaction. This can be done using a chemical reagent or distillation when the compounds of interest boil above 100°C. Among the materials able to eliminate water are desiccants such as anhydrous $MgSO_4$, molecular sieves, or substances that react with water such as CaC_2, $(CH_3)_2C(OCH_3)_2$ (2,2-dimethoxypropane), and even an appropriately chosen acid anhydride that reacts faster with water than with the reacting alcohol. The derivatization also may be performed in the presence of $SOCl_2$ (thionyl chloride), which reacts with the water assisting in its removal, and when present in excess, may react with the alcohols forming alkyl chlorides or with the acids forming acyl chlorides. Chloride is a better leaving group in a nucleophilic alkylation reaction, and the efficiency of alkylation increases. Acids also can be esterified using a mixture of an alcohol and an acyl halide.

One procedure for the formation of esters with less active organic acids applies the addition of dicyclohexylcarbodiimide (DCCI) in the derivatization process, to facilitate esterification. The reaction can be performed by adding to the acids that need to be analyzed the appropriate alcohol and DCCI usually in a solvent such as pyridine. Dicyclohexylurea, which is formed in the reaction, is not soluble in pyridine and can be separated. Besides DCCI, other carbodiimides can be used in the reaction of acids and alcohols. Among these are carbonyldiimidazole (CDI), 6-chloro-1-p-chlorobenzensulfonyloxybenzotriazole (CCBBT), 1-(3-dimethylaminopropyl)-3-ethylcarbodiimide (EDAC), etc. Also, 2-chloro-1-methylpyridinium iodide, 2,4,6-triisopropylbenzenesulfonyl chloride, trialkyloxonium fluoroborate, etc. can be used to facilitate esterification.

Transesterification is another technique applicable for obtaining certain alkyl derivatives of acids (or acyl derivatives of alcohols). The reaction can be written as follows:

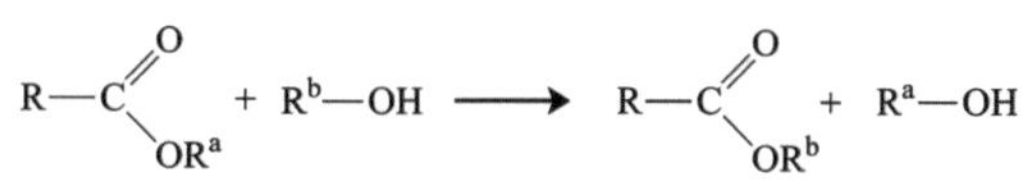

(15)

Transesterification can be catalyzed by acids (or Lewis acids) such as HCl, BF_3, and H_2SO_4 or by bases such as CH_3OK, CH_3ONa, or C_4H_9ONa. The basic catalysts are commonly used for the methanolysis of triglycerides, followed by the analysis of the fatty acid methyl esters using GC or GC/MS [16].

A special alkylation can be achieved online during the heating in the injection port of a gas chromatograph using tertraalkylammonium hydroxides or alkylarylammonium hydroxides. Tetramethylammonium hydroxide (TMAH) is the most common reagents of this type. The reaction takes place as follows (Δ indicates heating):

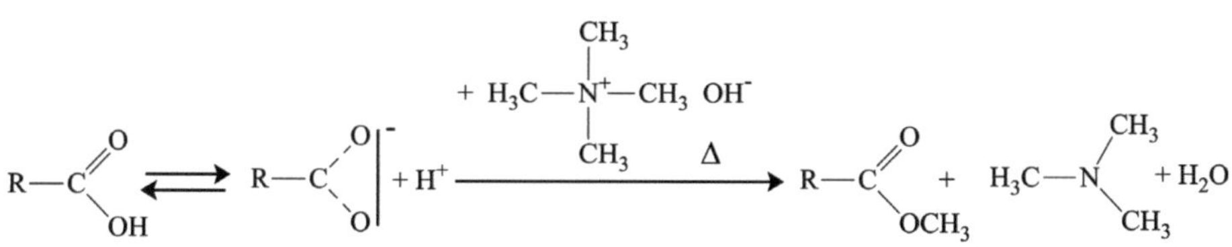

(16)

Numerous other reactive compounds may be used for replacing active hydrogens in specific compounds. For example, epoxides, aziridines, and episulfides react easily with compounds with active hydrogens. Formation of a second group containing an active hydrogen may preclude the use of such reagents for analytical purposes.

Besides the desired derivatives, certain unexpected compounds that can be considered artifacts for the particular analysis can also be formed in alkylation reactions. The artifacts may be formed from unexpected interactions of the reagent with the analyte or may be a result of undesired effects of the catalysts or medium used for derivatization. In some cases, the control of the alkylation process may be difficult. Longer or shorter reaction times or intervals between derivatization and analysis may lead to errors, even when an internal standard is used for quantitation.

One common case of artifact formation occurs during the reaction with compounds containing O-acyl or N-acyl groups, such as previously acylated carbohydrates, glycolipids, or glycoproteins, in particular when the reaction is done with short-chain alkyl bromides or iodides. When the OH groups of different sugars or NH_2 groups of amino sugars were already protected with acyl groups, it was noted that, depending on the catalyst and the chosen medium, these acyl groups can be replaced by alkyl groups, or they may migrate from one position (such as C1) to other positions.

Oxidation is another common side reaction when using Ag_2O as a catalyst. The oxidation effect of Ag_2O can be seen on free sugars as well as when attempting to permethylate peptides. Sulfhydryl groups are particularly sensitive to oxidation with Ag_2O as a catalyst. The use of methylsulfinyl carbanion as a methylating reagent may also produce undesired side reactions with certain esters generating methylsulfinylketones. Also, strong alkylating reagents may produce undesired artifacts by unexpected alkylations.

The derivatization with the purpose of obtaining aryl derivatives is similar in many respects to the alkylation reaction. The reaction takes place with compounds containing active hydrogens. Simple aryl halides are generally resistant to be attacked by nucleophiles and do not react similar to alkyl halides. This low reactivity can be significantly increased by changes in the structure of aryl halide or in the reaction conditions. The nucleophilic displacement can become very rapid when the aryl halide is substituted with electron attracting groups such as NO_2.

8. Silylation reactions

Silylation is the chemical reaction of replacing a reactive hydrogen atom in OH, COOH, SH, NH, CONH, POH, SOH, or enolisable carbonyl with a silyl group, most frequently with trimethylsilyl (TMS). A large number of analytical methods involve silylation applied to alcohols including carbohydrates [17], phenols [18], amines, sterols [19], etc. The purpose of silylation in chromatography is mainly to reduce the polarity of the analyte, increase its stability, and improve the GC behavior. The differences in the mass spectra of the silylated compounds as compared to the initial analyte may also be an advantage for detectability. However, the mass spectra of many silylated compounds may not be available in common mass spectral libraries. Also, the silylated compounds plus the commonly present excess of silylating reagent may deteriorate some types of stationary phases such as that of Carbowax (polyethylene glycol)-type columns, and for this reason, their separation cannot be done on such columns.

Silylation can be performed on specific analytes or directly on complex samples such as a plant material (see, e.g., [12]). The silylating agent and the solvent can play the double role of extractant and silylating reagent. Many publications describe the use of silylation reactions for analytical purposes (e.g., [1, 5, 20]). The reaction of an analyte Y:H with the formation of a TMS derivative can be written as follows:

$$\mathrm{Y{:}H} + \mathrm{H_3C{-}Si(CH_3)_2{-}X} \longrightarrow \mathrm{H_3C{-}Si(CH_3)_2{-}Y} + \mathrm{HX} \qquad (17)$$

The molecular weight for TMS is 73.047 calculated considering in the elemental composition of only the masses of the most abundant isotope. Numerous reagents have been synthesized to be used in silylations. Various aprotic solvents can be used as medium for silylation. The analysis can be focused on one analyte or on a mixture of analytes. The main factors contributing to the increase of the efficiency and the rate of the silylation reaction are the silyl donor ability of the reagent and the ease of silylation of different functional groups in the analyte. The solvent (or mixture of solvents) used as a medium and the compounds present or added in the silylation medium may also play a role for silylation efficiency. The reagent excess is sometimes important for displacing the equilibrium in the desired direction, and usually an excess up to ten times larger than stoichiometrically needed is used for silylation. Temperature also increases reaction rate, as expected, and heating of the sample with the reagents at temperatures around 70°C for 15 to 30 min is common. Some reagents used for trimethylsilylation are shown in **Figure 8** [14].

The approximate order of the increasing silyl donor ability for the reagents shown in **Figure 8** is HMDS < TMCS < MSA < TMSA < TMSDEA < TMSDMA < MSTFA < BSA < BSTFA < TMSI. This order may be different on particular substrates where other reagents or reagent mixtures may be more reactive.

Silylation reagents can be used pure or in mixtures of two or even three reagents. The reagent mixtures may provide a more efficient silylation for specific compounds. For example, silylation of 3,4-dimethoxyphenylethylamine with BSA leads to the substitution of only one active hydrogen in the NH_2 group, while the silylation with BSA in the presence of 5% TMCS produces silylation of both hydrogens in the NH_2 [21]. A common silylating mixture is BSTFA with 1% TMCS.

One of the determining factors regarding the silylation efficiency is the nature of the molecule Y:H that is being silylated (the analyte) and plays a crucial role in the choice of the derivatization conditions. It was noticed experimentally that the decreasing ease of silylation follows approximately the order shown in **Table 2**.

In general, the silylation of OH and COOH groups takes place with better results than that of NH_2, CONH, or NH groups. Excellent results are obtained, for example, for the analysis of phenols after silylation [19]. A chromatogram of a solution containing a mixture of phenols at concentrations between 2.0 and 2.5 μg/mL in DMF, derivatized with BSTFA, separated on a BPX-5 chromatographic column (SGE Anal. Sci.), followed by MS analysis in single-ion monitoring (SIM) mode is shown in **Figure 9**. Details regarding the analyzed phenols are given in **Table 3**.

Besides organic active hydrogens, several inorganic compounds with active hydrogens can also react with silylating reagents. Among these are H_2O, H_2O_2, and strong inorganic acids. Also, some salts of the acids may be silylated. The reaction of silylating reagents with water imposes that water should be at the low level in the matrix or the solution of the analytes. The reaction with water takes place as follows:

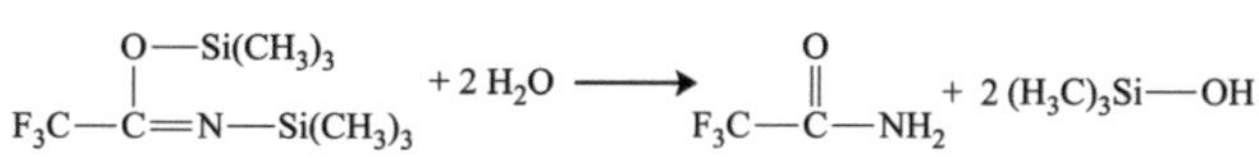
(18)

In many solvents used as medium for derivatization, the trimethylsilanol formed in the reaction with water is separated as a distinct layer of solvent. The formation of two layers impedes a proper sampling of the derivatized material in the GC/MS instrument. In addition to that, the presence of an excess of water suppresses the derivatization of other compounds. The silylation is not recommended on samples with a water content higher than about 10%.

The silylation reaction is commonly performed in a solvent that does not have active hydrogens. The most commonly used solvents as a medium for silylation are dimethylformamide (DMF), pyridine, and acetonitrile. The main role of the solvent is to dissolve the analyte and the reagents. The by-product HX of silylation shown in reaction (17) can be an acid, a base, or a neutral compound. As examples, for TMCS the by-product is HCl, for HMDS the

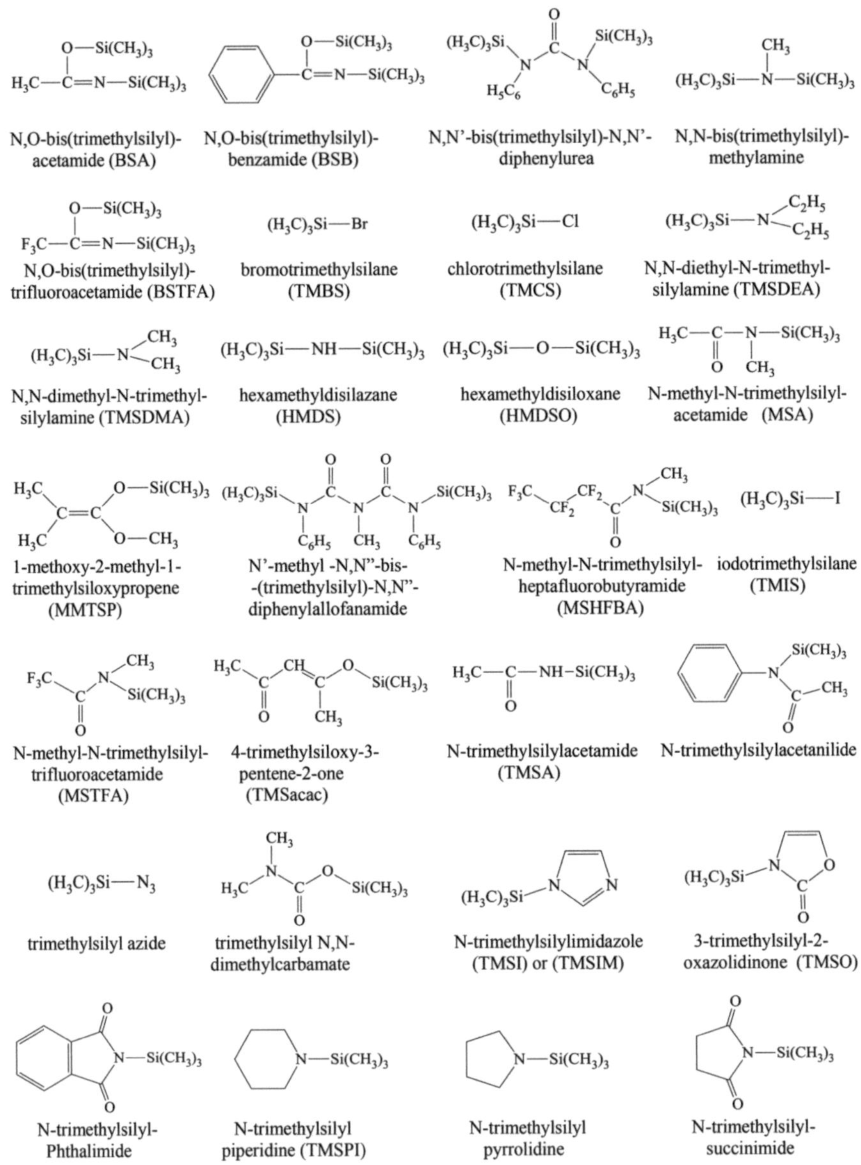

Figure 8.
Some reagents used for trimethylsilylation.

by-product is NH_3, for BSTFA the by-product is N-TMS-trifluoroacetamide, and for TMSI the by-product is imidazole. When the silylation reagent generates an acid as a by-product of the reaction, this may interfere with the silylation. For this reason, silylation can be promoted by any acid acceptor used as solvent or present in the solvent. Among such solvents are pyridine, triethylamine, and to a lower extent DMF. They can be used as both solvents and acid acceptors. Mixtures of solvents are commonly used for both enhancing solubility and promoting silylation. For example, formamide in the presence of pyridine may react with an acidic by-product generating CO and an ammonium salt. The addition of basic compounds to the silylation reaction may also influence the

	Compound	Functional group	Decreasing reactivity
(1)	Primary alcohol	OH	↓
(2)	Secondary alcohol	OH	
(3)	Tertiary alcohol	OH	
(4)	Phenol	OH	
(5)	Thiophenol	SH	
(6)	Aliphatic acid	COOH	
(7)	Aromatic acid	COOH	
(8)	Primary amine	NH_2	
(9)	Thiol	SH	
(10)	Amide	$CONH_2$	
(11)	N-TMS amide	CONH-$Si(CH_3)_3$	
(12)	Secondary amine	NH	
(13)	Indole	NH	

Table 2.
Several functional groups that can be silylated (listed in the approximate order of decreasing ease of silylation).

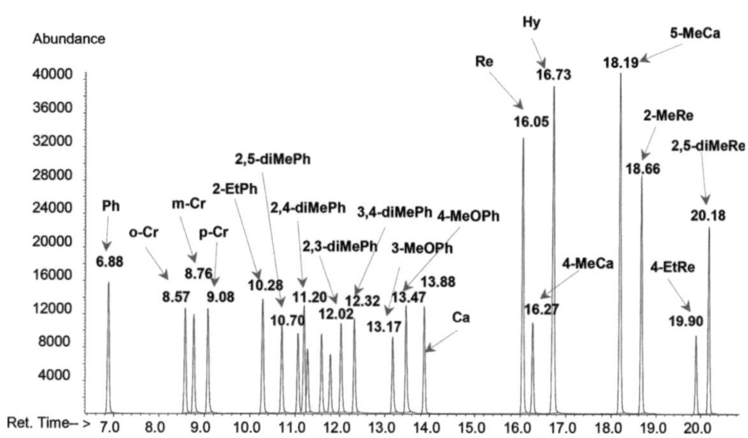

Figure 9.
Chromatogram of a set of phenol standards in DMF with the concentrations between 2.0 and 2.5 μg/mL derivatized with BSTFA, separated on a BPX-5 chromatographic column followed by MS analysis.

efficiency of the silylation. Also, some compounds may act as catalysts for silylation.

Although the TMS derivatives are by far the most commonly used in the derivatization for analytical purposes, other substituents in the silyl group can be used as reagents. Several such groups are indicated in **Figure 10**. The groups can be present in a variety of reagents connected to leaving groups "X-" such as Cl-, imidazolyl, F_3C-(CO)-N(CH_3)-, etc. For example, a common reagent containing *tert*-butyldimethylsilyl group is N-methyl-N-(*tert*-butyldimethylsilyl)-trifluoroacetamide (MTBSTFA), which has the following structure:

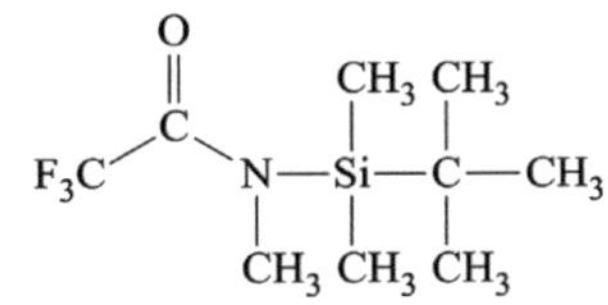

The use of different groups than TMS may serve different purposes. For example, a fluorinated or brominated group may enhance significantly the detection sensitivity when using ECD or NCI-MS. Also, the stability toward hydrolysis of compounds silylated with different groups than TMS may be higher, and such silylation can be advantageous. This is, for example, the case of *tert*-butyldimethylsilyl group that is typically more stable to hydrolysis than trimethylsilyl.

As an example, silylation of amino acids with MTBSTFA is commonly used [22, 23], and it is preferred to the silylation generating TMS derivatives. The chromatogram of a set of amino acid standards with the concentration of 0.05 μmol/mL derivatized with MTBSTFA and separated on a DB-5MS chromatographic column (from Agilent) followed by MS analysis is shown in **Figure 11**. Details regarding the analyzed amino acids are given in **Table 4**.

In most situations, silylation generates only the desired derivatives. However, there are cases when the expected silylated compound is not formed, and either the silylation is not complete, or some compounds such as aldehydes, ketones, or esters with no obvious active hydrogen generate silylated compounds. Incomplete

No.	Compound	Ret. time	m/z	Abrrev.	No.	Compound	Ret. time	m/z	Abrrev.
(1)	Phenol	6.88	166	Ph	(14)	3,4-Dimethylphenol	12.32	194	3,4-diMePh
(2)	o-Cresol	8.57	180	o-Cr	(15)	3-Methoxyphenol	13.17	196	3-MeOPh
(3)	m-Cresol	8.76	180	m-Cr	(16)	4-Methoxyphenol	13.47	196	4-MeOPh
(4)	p-Cresol	9.08	180	p-Cr	(17)	Catechol	13.88	254	Ca
(5)	2-Ethylphenol	10.28	194	2-EtPh	(18)	Resorcinol	16.05	254	Re
(6)	2,5-Dimethylphenol	10.70	194	2,5-diMePh	(19)	4-Methylcatechol	16.27	268	4-MeCa
(7)	3,5-Dimethylphenol	11.07	194	3,5-diMePh	(20)	Hydroquinone	16.73	254	Hy
(8)	2,4-Dimethylphenol	11.20	194	2,4 diMePh	(21)	3-Methylcatechol	16.71	268	3-MeCa
(9)	2-Methoxyphenol	11.28	196	2-MeOPh	(22)	5-Methylresorcinol	18.19	268	5-MeCa
(10)	4-Ethylphenol	11.59	194	4-EtPh	(23)	2-Methylresorcinol	18.66	268	2-MeRe
(11)	4-Chlorophenol	11.71	185	4-ClPh	(24)	4-Ethylresorcinol	19.90	282	4-EtRe
(12)	2,6-Dimethylphenol	11.79	194	2,6-diMePh	(25)	2,5-Dimethylresorcinol	20.18	282	2,5-diMeRe
(13)	2,3-Dimethylphenol	12.02	194	2,3-dimePh					

Table 3.
Details regarding the analyzed phenols with the chromatogram shown in Figure 9.

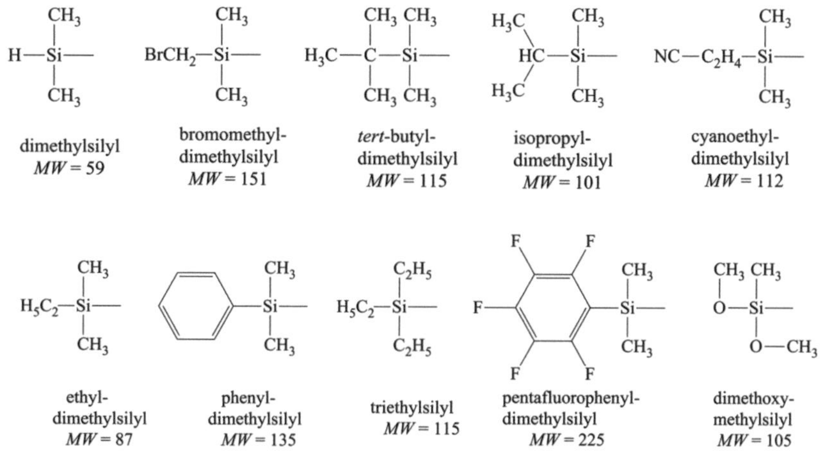

Figure 10.
Examples of silyl groups different from TMS used in silylation reagents.

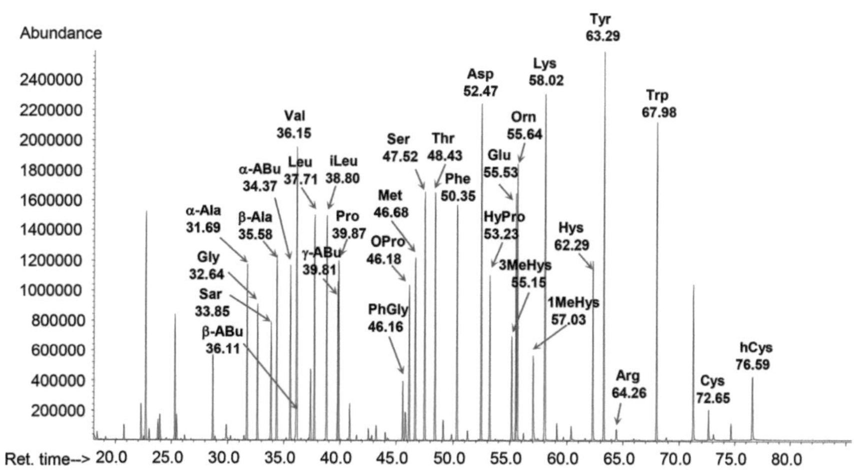

Figure 11.
Chromatogram of a set of amino acid standards with the concentration of 0.05 μmol/mL derivatized with MTBSTFA separated on a DB-5MS chromatographic column.

silylation is usually the result of inappropriate reaction conditions. However, when compounds with multiple functionalities are silylated, it is possible to generate a variety of derivatized compounds, regardless of the intention to obtain fully silylated or partly silylated compounds.

In some cases, artifacts are formed due to the modification of the analyte under the influence of the reagents during derivatization. For example, when the silylation is done in basic or acidic conditions, the analytes that are sensitive to acidic or basic media may suffer unexpected transformations. The most frequent artifacts with compounds not containing obvious active hydrogens occur with aldehydes. Some aldehydes are able to undergo two types of chemical reactions with formation of OH groups, namely, enolization and acetal formation in the presence of water. The OH groups formed in this manner react with different silylating reagents and give the corresponding silylated products. Although the enolization or the acetal formation is negligible for the initial aldehyde, the reactions may be significantly displaced toward the formation of the silylated compounds of the enol or of the acetal. Artifacts can also be generated when the reaction is allowed to continue for an

Peak No.	Amino acid	Abbrev.	MW	Formula + x TBDMS	MW + x TBDMS	Charact. ion	Ret. time
(1)	α-Alanine	α-Ala	89.09	$C_{15}H_{35}NO_2Si_2$	317	260	31.69
(2)	Glycine	Gly	75.07	$C_{14}H_{33}NO_2Si_2$	303	246	32.63
(3)	Sarcosine	Sar	89.09	$C_{15}H_{35}NO_2Si_2$	317	260	33.85
(4)	α-Amino-n-butyric acid	α-ABu	103.10	$C_{16}H_{37}NO_2Si_2$	331	274	34.36
(5)	β-Alanine	β-Ala	89.09	$C_{15}H_{35}NO_2Si_2$	317	260	35.58
(6)	Urea		60.06	$C_{13}H_{32}N_2OSi_2$	288	231	36.01
(7)	β-Aminoisobutyric acid	β-ABu	103.10	$C_{16}H_{37}NO_2Si_2$	331	274	36.11
(8)	Valine	Val	117.15	$C_{17}H_{39}NO_2Si_2$	345	186	36.15
(9)	Leucine	Leu	131.17	$C_{18}H_{41}NO_2Si_2$	359	200	37.71
(10)	Norleucine		131.17	$C_{18}H_{41}NO_2Si_2$	359	200	38.8
(11)	Isoleucine	iLeu	131.17	$C_{18}H_{41}NO_2Si_2$	359	200	38.8
(12)	γ-Aminobutyric acid	γ-ABu	103.10	$C_{16}H_{37}NO_2Si_2$	331	274	39.79
(13)	Proline	Pro	115.13	$C_{17}H_{37}NO_2Si_2$	343	184	39.87
(14)	2-Phenylglycine	PhGly	151.17	$C_{20}H_{37}NO_2Si_2$	379	220	46.16
(15)	5-Oxoproline	oPro	129.13	$C_{17}H_{35}NO_3Si_2$	357	300	46.18
(16)	Methionine	Met	149.20	$C_{17}H_{39}NO_2SSi_2$	377	320	46.68
(17)	Serine	Ser	105.09	$C_{21}H_{49}NO_3Si_3$	447	390	47.52
(18)	Threonine	Thr	119.12	$C_{22}H_{51}NO_3Si_3$	461	404	48.43
(19)	Phenylalanine	Phe	165.19	$C_{21}H_{39}NO_2Si_2$	393	336	50.35
(20)	Aspartic acid	Asp	133.10	$C_{22}H_{49}NO_4Si_3$	475	418	52.47
(21)	Hydroxyproline	HyPro	131.13	$C_{23}H_{51}NO_3Si_3$	473	314	53.23
(22)	3-Methyl-L-histidine	3MeHys	169.20	$C_{19}H_{39}N_3O_2Si_2$	397	340	55.15
(23)	Glutamic acid	Glu	147.13	$C_{23}H_{51}NO_4Si_3$	489	432	55.53
(24)	Ornithine	Orn	132.20	$C_{23}H_{54}N_2O_2Si_3$	474	286	55.64
(25)	1-Methyl-L-histidine	1MeHys	169.20	$C_{19}H_{39}N_3O_2Si_2$	397	302	57.03
(26)	Lysine	Lys	146.19	$C_{24}H_{56}N_2O_2Si_3$	488	300	58.02
(27)	α-Aminoadipic acid		161.20	$C_{24}H_{53}NO_4Si_3$	503	446	58.06
(28)	Histidine	Hys	155.16	$C_{24}H_{51}N_3O_2Si_3$	497	440	62.29
(29)	Tyrosine	Tyr	181.19	$C_{27}H_{53}NO_3Si_3$	523	302	63.29
(30)	Arginine	Arg	174.20	$C_{24}H_{56}N_4O_2Si_3$	516	144	64.26
(31)	Tryptophan	Trp	204.22	$C_{29}H_{54}N_2O_2Si_3$	546	244	67.98
(32)	Cystine	Cys	240.30	$C_{28}H_{64}N_2O_4S_2Si_4$	668	348	72.65
(33)	Homocystine	hCys	268.30	$C_{32}H_{72}N_2O_4S_2Si_4$	724	362	76.59

Table 4.
Details regarding the analyzed amino acids with the chromatogram shown in Figure 11.

extended period of time. Other uncommon reactions with a specific silylation reagent and analyte may occur. An example of an uncommon reaction is the ring opening of flavanones.

9. Acylation reactions

The formation of acyl derivatives is applied for replacing the active hydrogens from an analyte in functionalities such as OH, SH, NH [11, 24], CONH, etc. The acylation is also used for reducing polarity and improving the behavior of the analytes in the chromatographic column. Acylation may confer a better volatility of the analytes, although not as marked as for silylation or methylation. Only the derivatization with acetyl groups or with fluorinated acyl groups (not heavier than heptafluorobutyryl) improves volatility, while other heavier acyl groups are not suitable for this purpose. Acetylation, for example, can be used for compounds such as monosaccharides and amino acids to allow GC analysis. The detectability improvement on the other hand is a very common purpose for acylation. Acylation with fluorinated compounds is frequently used for enhancing detectability in GC with ECD or NCI-MS detection. Other uses of acylation include the enhancement of separation of chiral compounds, etc.

Most acylation reactions are nucleophilic substitutions where the analyte is a nucleophile (Y:, Y:H, Y:-) reacting with the acylating reagent RCO-X that contains a leaving group X and an acyl group RCO: as shown in the following reaction:

$$\mathrm{Y{:}H} + \mathrm{R{-}\overset{\overset{\displaystyle O}{\|}}{C}{-}X} \rightleftarrows \mathrm{R{-}\overset{\overset{\displaystyle O}{\|}}{C}{-}Y} + \mathrm{X{:}H} \tag{19}$$

Some common acyl groups present in acylation reagents are indicated in **Table 5**.

As shown in **Table 5**, the acyl groups in the reagent can be attached to various "X" groups. One such group is OH and among the acylating reagents are some free acids. When nucleophile is an alcohol, the reaction is known as esterification and has been discussed in Section 7. The acylation with acids can be applied besides alcohols to certain thiols, phenols, amines, etc. and can be written as follows:

$$\mathrm{Y:H + R - COOH \rightarrow R - COY + H_2O} \tag{20}$$

The reaction can be displaced toward the formation of the acyl derivatives by eliminating the water using compounds such as anhydrous $MgSO_4$, molecular sieve, or substances that react with water such as CaC_2, or $(CH_3)_2C(OCH_3)_2$. Dicyclohexylcarbodiimide (DCCI) also is used for modifying the yield of the desired product. The reaction with reagents containing a carboxylic acid reactive group also can be done in the presence of 2,4,6-trichlorobenzoyl chloride or with various sulfonyl chlorides such as 2,4,6-triisopropyl-benzenesulfonyl chloride or 2,4,6-trimethylbenzenesulfonyl chloride. The reaction of amines with acids can be displaced toward the formation of the amides using a peptide coupling reagent such as benzotriazol-1-yl-oxy-tris(dimethyl-amino)-phosphonium hexafluorophosphate (BOP), diethyl cyanophosphonate, O-benzotriazol-1-yl-N,N,N′,N′-bis (tetramethylene)uronium hexafluorophosphate, 2,2′-dipyridyl disulfide + triphenylphosphine, 1-(3-dimethylaminopropyl)-3-ethylcarbodiimide (EDAC), etc.

Common acylating reagents are acyl halides such as chlorides or bromides, which are reactive compounds suitable for acylation. The reaction of an acyl chloride with an amine, for example, takes place as follows:

Group	Group structure	Mass of the group	Example of reagents	Common analytes
Formyl	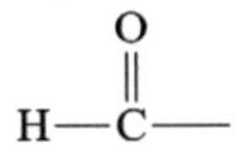	29	Formic acid	Steroids
Acetyl	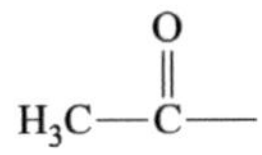	43	Acetyl chloride, acetic anhydride	Alcohols
Trifluoroacetyl	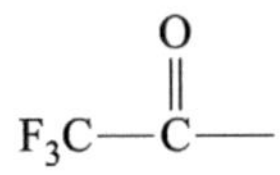	97	N-Methyl-bis (trifluoroacetamide), bis(trifluoroacetamide), trifluoroacetic acid (TFA)	Alcohols
Propionyl	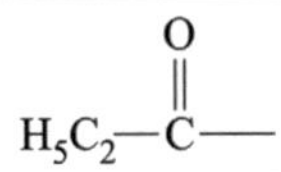	57	Propionic anhydride	Alcohols
Butyryl	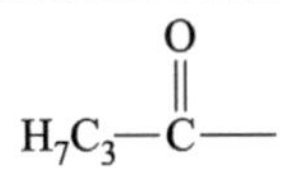	71	Butyric anhydride	Alcohols
2,2-Dimethyl-propionyl-(pivaloyl)	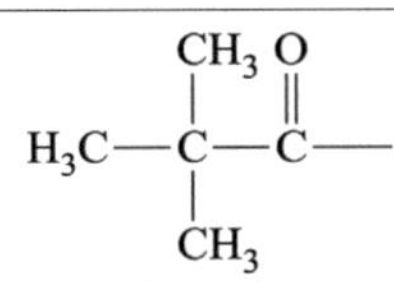	85	Pivaloyl chloride, pivalic anhydride	Amino acids
Pentafluoro-propionyl	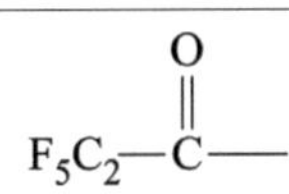	147	Pentafluoropropionic anhydride (PFPA)	Alcohols, amines, amino acids
Heptafluorobutyryl	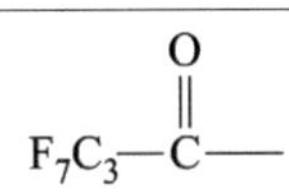	197	Heptafluorobutyric anhydride (HFBA), heptafluorobuty rylimidazole	Alcohols, amines, amino acids
Trichloroacetyl	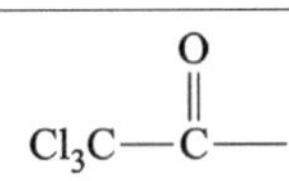	145	Trichloroacetic anhydride	Alcohols
Pentafluorobenzoyl	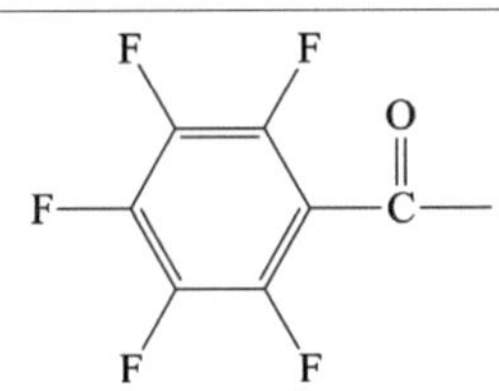	195	Pentafluorobenzoyl chloride, pentafluorobenzoyl-imidazole	Alcohols, amides
(Pentafluorophenyl)-acetyl	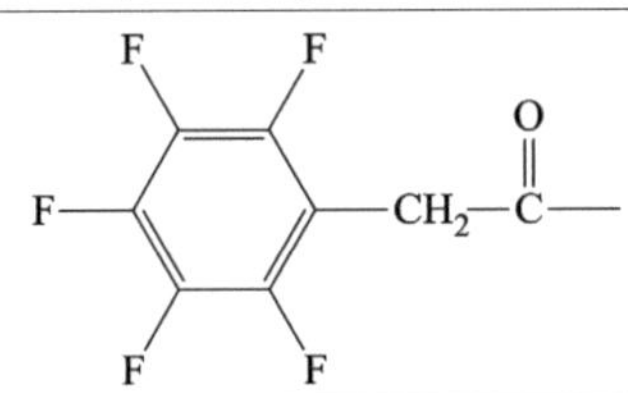	209	(Pentafluorophenyl)-acetyl chloride	Alcohols, amides
(Pentafluorophenoxy)-acetyl	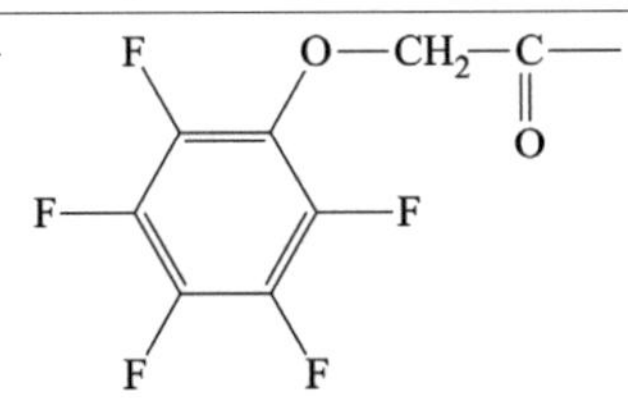	225	(Pentafluorophenoxy)-acetyl chloride	Alcohols

Table 5.
Some common groups present in acylating reagents used in derivatizations for GC analysis [14].

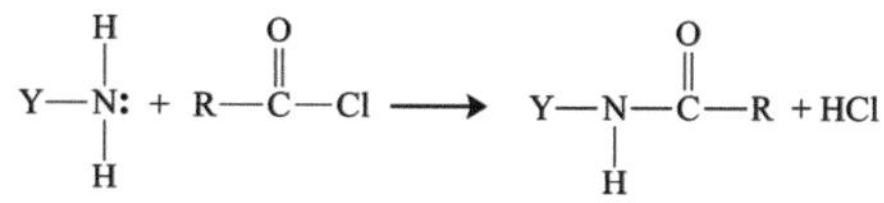

(21)

Since the reactivity of amides is lower than that of amines, the second hydrogen in the amine is more difficult to replace. Also, steric hindrance may negatively influence the reaction. The generation of a strong acid such as HCl is a disadvantage in the reaction with acyl halides, and usually the acid should be removed either by adding basic compounds such as Na_2CO_3 or $MgCO_3$ or using pyridine as the reaction medium. The high reactivity of acyl halides is used for the acylation of compounds with less reactive hydrogens. Certain carbonyl cyanides react similarly to acyl chlorides.

The disadvantage of generating a strong inorganic acid in the acylation with acyl halides also can be avoided by having, instead of the acyl halide, an anhydride. The reaction of Y:H with an anhydride takes place as follows:

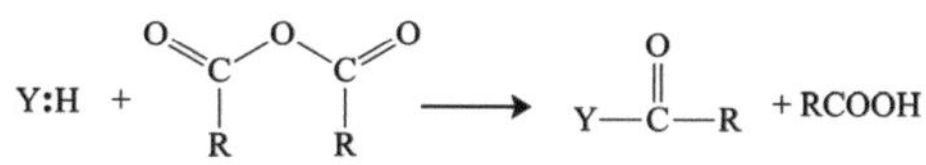

(22)

The acid resulting together with the acylated compound is not a strong acid such as HCl. The anhydrides of trifluoroacetic acid (TFA), pentafluoropropionic anhydride (PFPA), and heptafluorobutyric (HFBA) acids are commonly used for derivatization of alcohols, phenols, amines, etc., with the purpose of enhancing detectability (by ECD or NCI-MS) and also for improving the chromatographic behavior (higher volatility, better thermal stability, better separation). The volatility of fluorinated compounds allows the GC applications. The reactivity of the perfluorinated anhydrides increases in the order HFBA < PFPA < TFA. However, the differences are not significant. Once formed, the heptafluorobutyrates are more stable to hydrolysis than the trifluoroacetates. An inert solvent such as CH_2Cl_2, ether, ethyl acetate, acetone, tetrahydrofuran or in CH_3CN, etc. can be used as a medium for the reaction with perfluoroanhydrides. For the neutralization of the acids formed during derivatization, the basic compounds such as triethylamine, pyridine, or even solid $NaHCO_3$ can be utilized.

In order to avoid the formation of water or of a strong acid in the acylation reaction, certain amides such as N-methyl-bis(trifluoroacetamide), bis(trifluoroacetamide), or 2,2,2-trifluoro-N-methyl-N-(2,2,2-trifluoroacetyl)acetamide (MBTFA) can be used as reagents. Acylation of amines takes place at room temperature. Solvents such as CH_3CN, pyridine, DMSO, or THF can be used as a reaction medium:

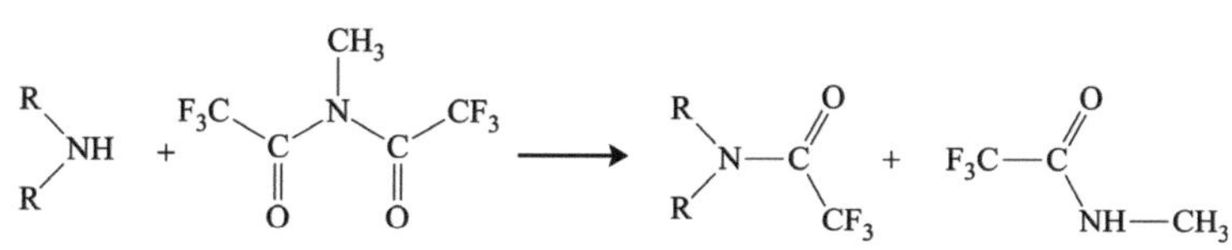

(23)

One other procedure successfully applied to obtain acyl derivatives is the use of acyl imidazoles as reagents. This class of compounds reacts with analytes containing alcohol, primary and secondary amino groups, or thiols. The reaction generates as a by-product imidazole:

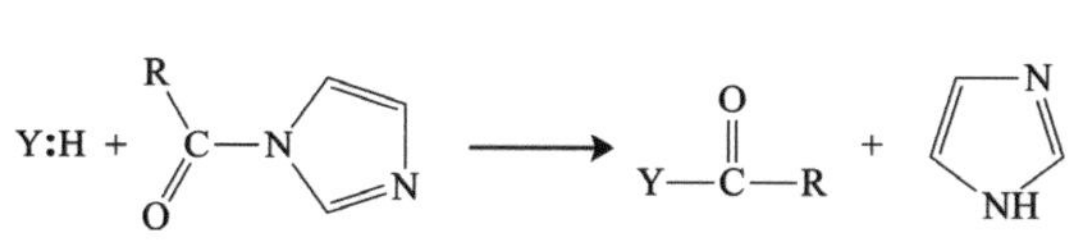

(24)

Succinimidyl esters also can be used for acylation purposes. Amines and the amino group in amino acids also can be acylated using urethane-protected α-amino acid-N-carboxyanhydrides or oxycarbonyl-amino acid-N-carboxyanhydrides. Alkylketenes and their dimers may be used for acylation.

A special type of acylation is that using chloroformates. Carbonic acid, O═C(OH)$_2$, can form amides, esters, halides, etc., due to the presence of two OH groups bonded to the CO group. Carbonic acid ester halides, also called chloroformates or chloroformate esters, with the formula R─O─C(═O)─X, where R is an alkyl or aryl group and X is F, Cl, Br, or I, can react with various compounds containing active hydrogens, such as acids [25], amines, alcohols, thiols, and amino acids. Amino acids, for example, in the presence of an alcohol in water form carbamate esters (urethanes) reacting as follows [26]:

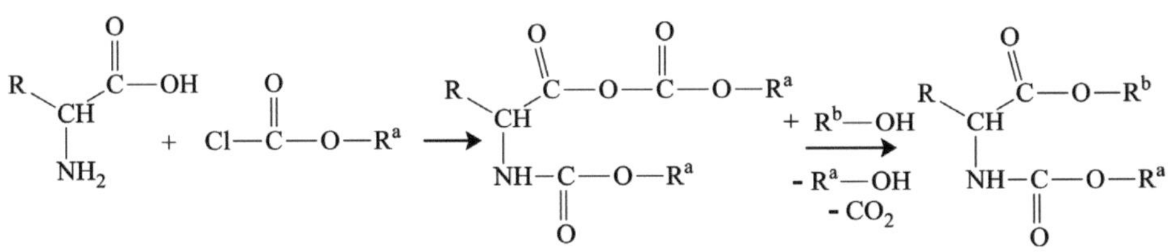

(25)

The formation in reaction (25) of the alcohol R^a–OH may lead to traces of a resulting compound with both substituted radicals being R^a. For this reason it is typically recommended to perform the reaction in the presence of an alcohol having the same radical as the chloroformate reagent ($R^a = R^b$). Chloroformates containing in the alkyl or aryl group halogen substituents are particularly reactive. Even tertiary amines can react with specific chloroformates, such as pentafluorobenzoyl chloroformate or with trichloroethyl chloroformate, by displacing an alkyl group connected to the nitrogen atom and forming the carbamate ester.

Similar in many respects to that of acyl derivatives R–CO–X are the reactions of sulfonyl derivatives R–SO$_2$–X. Sulfonyl halides are in general less reactive than halides of carboxylic acids. The reaction of a sulfonyl derivative may take place with alcohols, phenols, amines, etc. The reactivity toward the sulfonyl sulfur is $RNH_2 > CH_3COOR > H_2O > ROH$.

High reactivity toward active hydrogens in alcohols, amines, etc. can also be achieved using reagents with other functionalities. These functionalities include isocyanates, isothiocyanates, carbonyl azides, etc. These reactions can be seen as a replacement of an active hydrogen with a CO-R group or CS-R group as it occurs in other acylations.

10. Other derivatization reactions

A variety of other derivatization reactions are reported in the literature (see, e.g., [1]) and used for GC and GC/MS analyses. Among these are the addition to hetero multiple bonds in functional groups such as C═O, C═S, C═N, or C≡N. Many such reactions are additions to multiple bonds. Such reactions are, for example, the additions to the C═O groups in aldehydes and ketones. Reagents containing active hydrogens in groups such as NH$_2$, OH, H$_2$N-NH-, etc. can react, for example, with aldehydes and ketones. Alcohols, for example, form hemiacetals or acetals with aldehydes and ketals with ketones, and although most of such compounds are not stable enough to be suitable for derivatization, cyclic acetals and ketals may be stable and used for analytical purposes. A common reaction of carbonyl compounds is with amines. The initial addition reaction usually continues with water elimination forming a substituted imine or a Schiff base. Similar to the reaction of amines is the reaction with substituted hydroxylamines or hydrazines. A typical reaction of derivatization of carbonyl compounds is that using dinitrophenylhydrazine (DNPH). The derivatized compound can be analyzed either by LC [27] or by GC/MS [28]. The reaction takes place as follows:

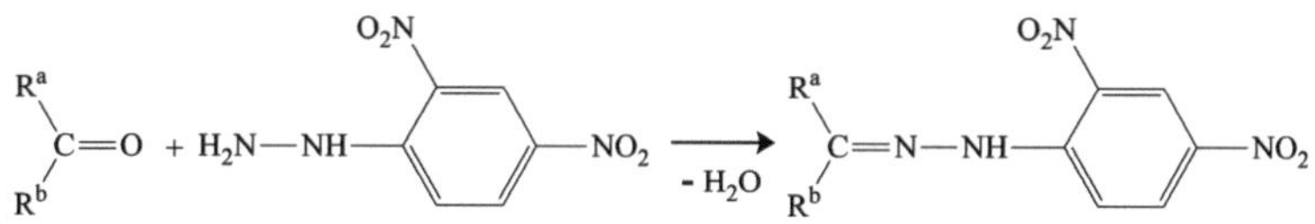

(26)

The groups R^a and R^b can be H or alkyl or various other substituents.

Another reagent that can be used for ketone derivatization is N-aminopiperidine in the presence of catalytic amounts of acetic acid. The resulting substituted hydrazone can be used in GC analysis:

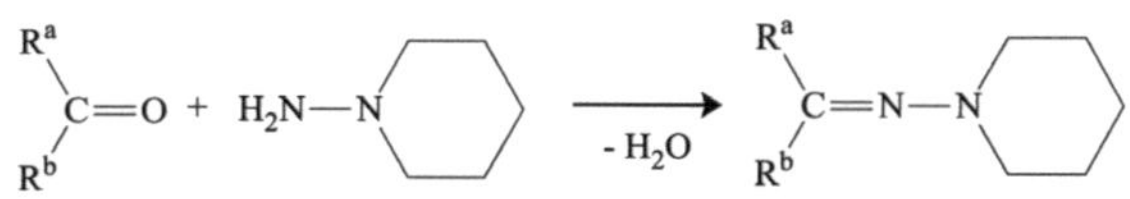

(27)

β-Diketones may react differently with hydrazines generating pyrazole derivatives as shown below:

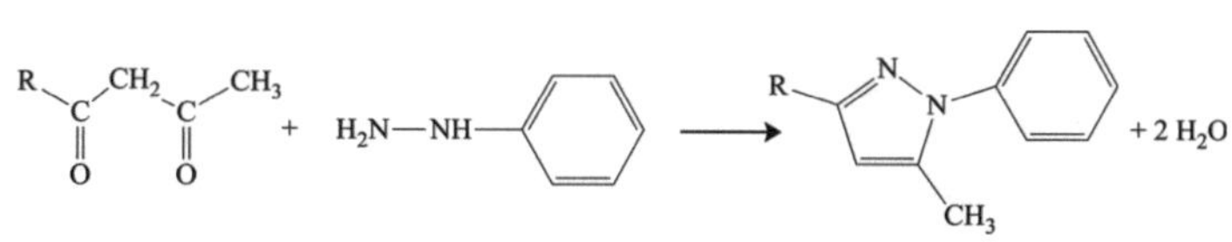

(28)

Several other classes of compounds similar to hydrazines react with the carbonyl compounds. Among these are hydrazones (NH_2—N=CR_2), hydrazides (NH_2NH-COR), and semicarbazide (NH_2NH-$CONH_2$). Hydroxylamines also react with carbonyl compounds forming oximes. Hydroxylamine itself, hydroxylamine hydrochloride (STOX® reagent), or derivatives such as H_2N-OSO_3H in a solvent like pyridine can be used in this reaction:

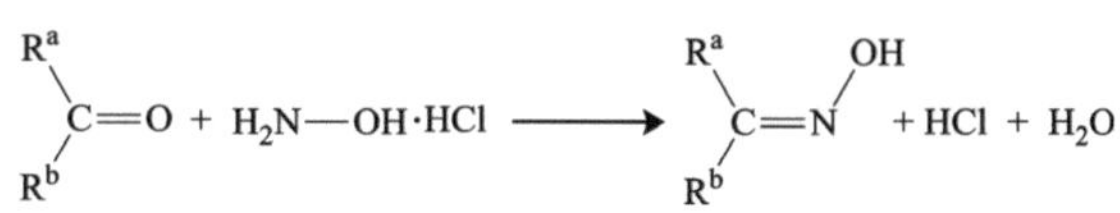

(29)

When the reaction is performed with hydroxylamine, the generated oxime contains an active hydrogen. This can be further derivatized, for example, by silylation in a reaction with a common silylation reagent.

For derivatization purposes other reagents can be used, such as substituted hydroxylamines like methoxyamine hydrochloride NH_2OCH_3•HCl (MOX® reagent) and O-(pentafluorobenzyl)-hydroxylamine hydrochloride (FLOROX® reagent). The reaction of a ketone or aldehyde with FLOROX is shown below:

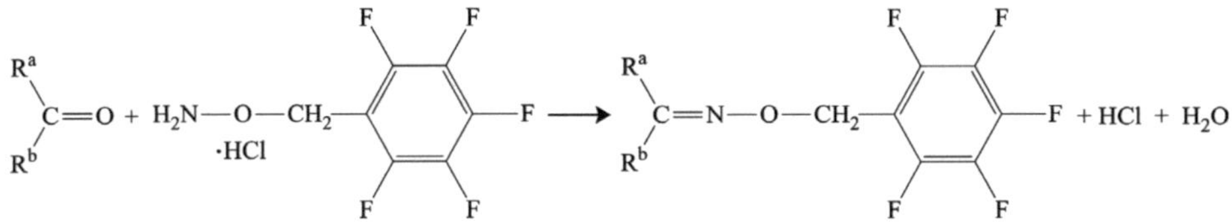

(30)

The oximes existing in *sin*- and *anti*- forms can produce double peaks in the GC chromatographic separations. To avoid this effect, oximes can be converted into nitriles when treated with acetic anhydride in the presence of CH_3COONa. This reaction was used in the derivatization of carbohydrates when a simultaneous acetylation takes place (Wohl degradation). The reaction can be written as follows:

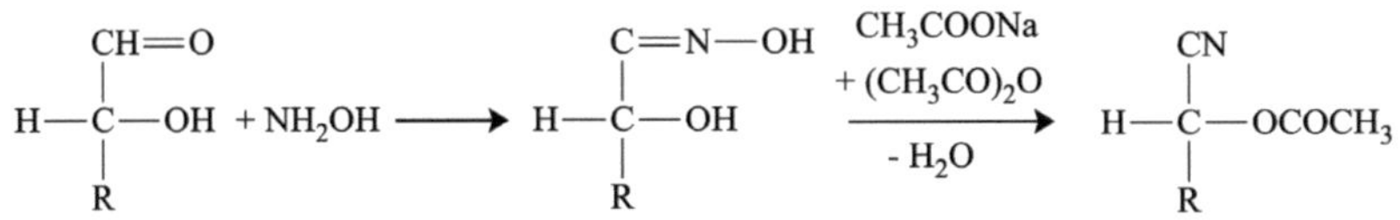

(31)

The transformation of the oximes into nitriles generates one single compound from the two (syn- and anti-) isomers and can be used to simplify the chromatograms of sugars derivatized as oximes.

Alcohols, amines, and thiols also can react at other hetero multiple bonds leading to analytical applications. This addition may occur at the isocyanates (—N=C=O), —C=O group in an amide, at a nitrile, at CS_2, or at other groups. One example is the addition under special conditions of alcohols to dimethylformamide. The resulting acetals are very reactive and are used themselves as reagents, as shown previously for N,N-dimethylformamide dimethyl acetal (see reaction 12). Another example is the reaction of CS_2 with alcohols in the presence of a base, leading to the formation of xanthates. Amines also react with CS_2, and the formed isothiocyanate can be analyzed using GC analysis. The reaction takes place as follows:

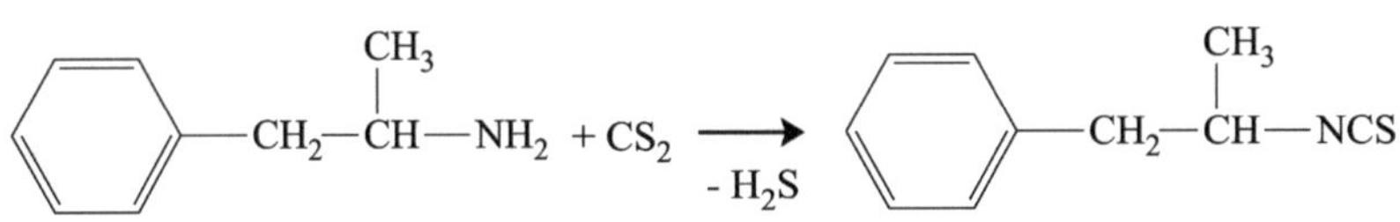

(32)

Formation of new cycles from noncyclic compounds or replacement of old cycles with new ones that are more stable or have a desired property is also exploited in sample processing using derivatization. Epoxides, for example, can be formed in the reaction of a compound with a carbon–carbon double bond and a peroxy acid. Among the peroxy acids more frequently used for the formation of epoxides are peracetic, performic, perbenzoic, trifluoroperacetic, and 3,5-dinitroperoxybenzoic acids. However, in this reaction a mixture of enantiomers is formed, as shown below for a *cis* olefin:

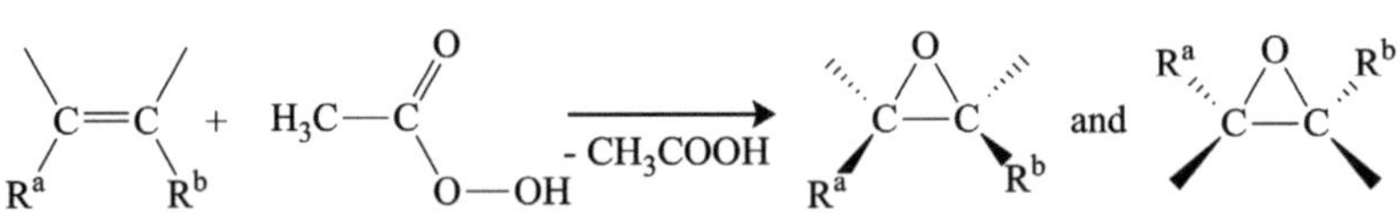

(33)

The separation of the epoxides may be easier to achieve than that of olefins, and this type of derivatization has been utilized, for example, for better separation of various *cis* and *trans* unsaturated fatty esters.

Another reaction with formation of new cycles is that of amino acids with phenyl isothiocyanate leading to a thiohydantoin derivative:

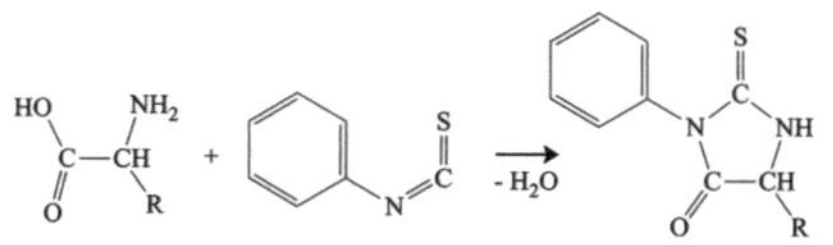
(34)

This reaction has been successfully used for the analysis of amino acids in proteins [29, 30]. *p*-Bromophenyl isothiocyanate has been used in a similar reaction.

A variety of aromatic cycles can be formed in reactions involving bifunctional compounds. Addition reactions to hetero multiple bonds in bifunctional molecules frequently lead to cyclic compounds. For example, formaldehyde can react with tryptophan or tryptamine generating a β-carboline derivative as follows:

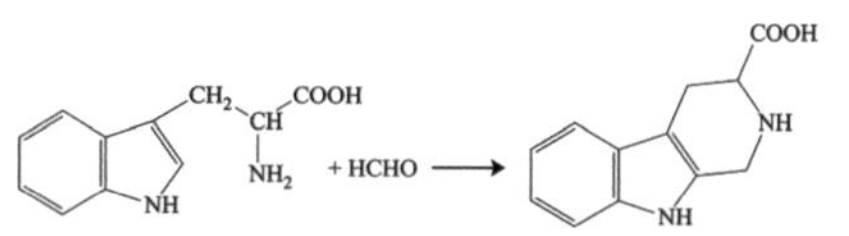
(35)

The new compound can be analyzed by GC, usually after further derivatization by silylation of the carboxyl group.

A typical reaction leading to pyrazoles is the reaction of hydrazines with diketones such as 2,4-pentandione (acetylacetone). For example, the reaction between hydrazine or methylhydrazine and acetylacetone takes place as follows:

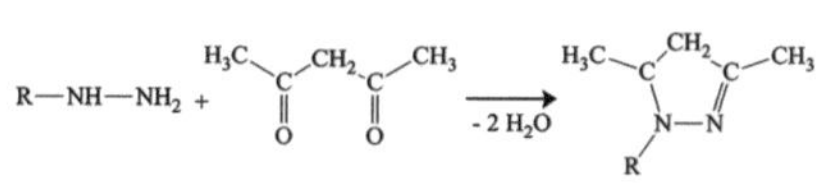
(36)

The resulting compound can be analyzed using a GC separation.

Activated carbonyl groups such as those in hexafluoroacetone are known to react with difunctional compounds. The reaction may take place with an amino acid as follows:

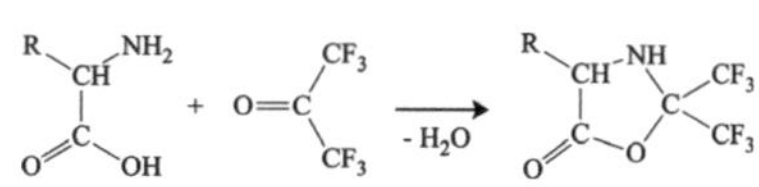
(37)

Amino acids can react with an activated anhydride such as trifluoroacetic anhydride (TFAA):

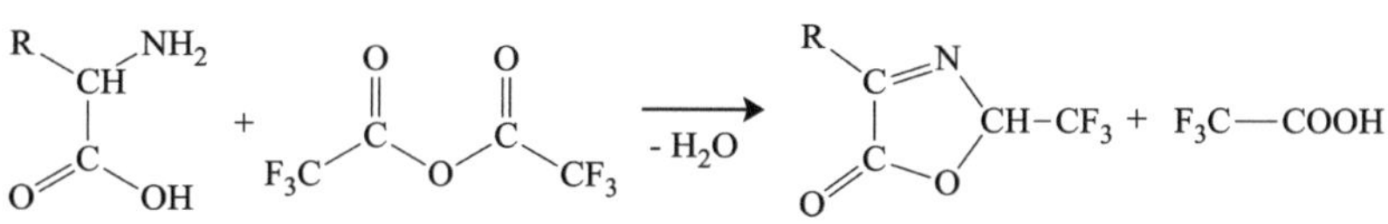
(38)

The reaction takes place by heating the amino acids with an excess of TFAA. The reaction mixture is then dissolved in ethyl acetate and analyzed by GC.

Numerous other types of derivatization reactions were used for making the analytes suitable for GC and GC/MS analyses. These include formation of various cyclic types of compounds such as azines, siliconides, boronates, etc., that are thermally stable and do not have polar hydrogens such that GC or GC/MS analysis is possible. In addition to reagents that add specific moieties to the analytes, oxidation and reduction were sometimes used for the analyte modification (see, e.g., [4]).

11. Derivatization reactions involving solid-phase reagents or derivatization on a solid support

Solid-phase reagents are polymeric materials with specific groups that are reactive and can be transferred to the analyte molecule producing derivatization. For an analyte of the form Y:H, the reaction with a solid-phase reagent can be written as follows:

$$\text{Y:H} + \text{R-Polymer} \rightarrow \text{R} - \text{Y} + \text{H-Polymer} \qquad (39)$$

Solid-phase reagents must work analogously to the corresponding small-molecule reagents containing the group R (a tag). Reagents that are insoluble in certain solvents at high concentrations can often provide a high ratio of analyte/substrate in a polymeric microenvironment that yields a high kinetic rate for the heterogeneous reaction.

A variety of materials can be used as solid support, such as specifically bound reagents on a silica support (used, e.g., for online derivatization in HPLC analysis), ion exchange resins, as well as other supports [31]. One example of solid-phase support that can produce derivatization is trifluoroacetyl nylon 6,6. This solid-phase reagent can be obtained from poly(hexamethylene adipamide) (nylon 6,6) and trifluoroacetyl anhydride. This solid-phase reagent can be used in amine derivatization in a reaction as follows:

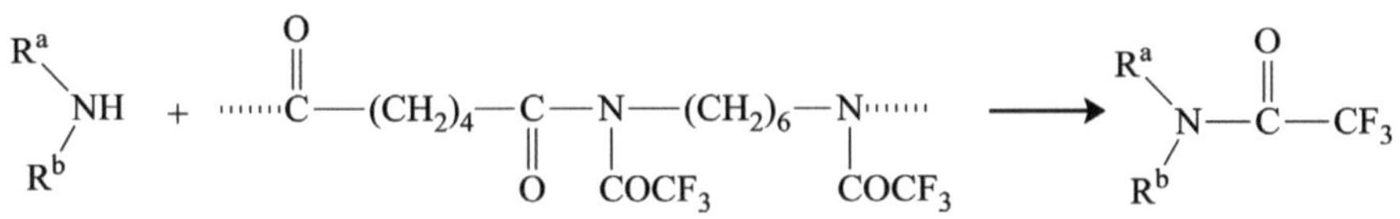

(40)

This derivatization of the amine is done by mixing the solid-phase reagent with a solution of amine solution in CH_3CN. Following derivatization, the solid-phase reagent is separated by centrifugation, and the solution is concentrated by evaporating part of the solvent and analyzed by GC (an amine internal standard must be used in this procedure). However, some such derivatizations require a long time of interaction between the solid-phase reagent and the analytes and found only limited applications.

(Another) alternative of derivatization of specific analytes is using the reaction between the reagent and the analyte both adsorbed on a solid support. This type of derivatization has been used, for example, in connection with a solid-phase microextraction (SPME) technique [32]. In this technique a reagent is initially adsorbed in the SPME fiber, followed by exposure to the analytes. The derivatized analytes are further desorbed in the injection port of the GC and analyzed using a detector such as MS. For example, formaldehyde from air can be analyzed using a polydimethylsiloxane (PDMS) fiber containing *o*-(2,3,4,5,6-pentafluorobenzyl)-hydroxylamine hydrochloride as a reagent. After exposing the fiber to the air sample contaminated with formaldehyde, the derivatization agent reacts with formaldehyde absorbed onto the coating forming an oxime. The oxime is thermally desorbed in a GC injector port and analyzed by GC with ECD [33].

Author details

Serban C. Moldoveanu[1*] and Victor David[2]

1 R.J. Reynolds Tobacco Co., Winston-Salem, NC, USA

2 University of Bucharest, Bucharest, Romania

*Address all correspondence to: moldovs@rjrt.com

References

[1] Knapp DR. Handbook of Analytical Derivatization Reactions. New York: J. Wiley; 1979

[2] Blau K, Halket J, editors. Handbook of Derivatives for Chromatography. Chichester: J. Wiley; 1993

[3] Toyo'oka T, editor. Modern Derivatization Methods for Separation Sciences. Chichester: J. Wiley; 1999

[4] Moldoveanu SC, David V. Sample Preparation in Chromatography. Amsterdam: Elsevier; 2002

[5] Pierce AE. Silylation of Organic Compounds. Rockford: Pierce Chem. Co.; 1982

[6] Moldoveanu SC. Evaluation of several minor alkaloid levels in tobacco with separation of (R)- and (S)-nornicotine. Beiträge zur Tabakforschung International. 2013;**25**: 649-659

[7] Ahuja S, editor. Chiral Separation Methods for Pharmaceutical and Biotechnological Products. Hoboken: John Wiley & Sons; 2011

[8] Hughes RO, Bronner WE, Smith ML. Detection of amphetamine and methamphetamine in urine by gas chromatography/mass spectrometry following derivatization with (−)-menthyl chloroformate. Journal of Analytical Toxicology. 1991;**15**:256-259

[9] Available from: https://www.activatedresearch.com/Public_Documents/Polyarc_Reactor_System_Product_Brochure.pdf

[10] Forehand J, Dooly GL, Moldoveanu SC. Analysis of polycyclic aromatic hydrocarbons, phenols and aromatic amines in particulate phase of cigarette smoke using simultaneous distillation and extraction as a sole sample clean-up step. Journal of Chromatography. A. 2000;**898**:111-1124

[11] Smith CJ, Dooly GL, Moldoveanu SC. New technique using solid-phase extraction for the analysis of aromatic amines in mainstream cigarette smoke. Journal of Chromatography. A. 2003; **991**:99-107

[12] Moldoveanu SC, Scott WA. Analysis of four pentacyclic triterpenoid acids in several bioactive botanicals with gas and liquid chromatography and mass spectrometry detection. Journal of Separation Science. 2016;**39**:324-332

[13] Held PK, White L, Pasquali M. Quantitative urine amino acid analysis using liquid chromatography tandem mass spectrometry and a TRAQ® reagents. Journal of Chromatography B. 2011;**879**:2695-2703

[14] Moldoveanu SC, David V. Modern Sample Preparation for Chromatography. Amsterdam: Elsevier; 2015

[15] Ciucanu I, Kerek F. A simple and rapid method for the permethylation of carbohydrates. Carbohydrate Research. 1984;**131**:209-217

[16] Bannon CD, Craske JD, Hai NT, Harper NL, O'Rourke KL. Analysis of fatty acid methyl esters with high accuracy and reliability II. Methylation of fats and oils with boron trifluoride methanol. Journal of Chromatography. 1982;**247**:63-69

[17] Sweeley CC, Bentley R, Makita M, Well WW. Gas-liquid chromatography of trimethylsilyl derivatives of sugars and related substances. Journal of the American Chemical Society. 1963;**85**: 2497-2507

[18] Moldoveanu SC, Kiser M. Gas chromatography/mass spectrometry

versus liquid chromatography/ fluorescence detection in the analysis of phenols in mainstream cigarette smoke. Journal of Chromatography A. 2007; **1141**:90-97

[19] Wells WW, Makita M. The quantitative analysis of fecal neutral sterols by gas-liquid chromatography. Analytical Biochemistry. 1962;**4**:204-212

[20] van Look G, Simchen G, Heberle J. Silylating Agents. Buchs: Fluka Chemie AG; 1995

[21] VandenHeuvel WJA. Gas-liquid chromatography of the trimethylsilyl derivatives of several amines of biological interest. Journal of Chromatography. 1968;**36**:354-358

[22] Mawhinney TP, Robinett RSR, Atalay A, Madson MA. Analysis of amino acids as their t-butyldimethylsilyl derivatives by gas-liquid chromatography and mass spectrometry. Journal of Chromatography. 1986;**358**:231-242

[23] Moldoveanu SC. Analysis of protein amino acids in tobacco using microwave digestion of plant material. Beiträge zur Tabakforschung International. 2005;**21**: 451-465

[24] Sasaki TA, Wilkins JM, Forehand JB, Moldoveanu SC. Analysis of heterocyclic amines in mainstream cigarette smoke using a new NCI GC-MS technique. Analytical Letters. 2001; **34**:1749-1761

[25] Husek P, Simek P, Matucha P. Smooth esterification of di- and tricarboxylic acids with methyl and ethyl chloroformates in gas chromatographic profiling of urinary acidic metabolites. Chromatographia. 2003;**58**:623-630

[26] Husek P. Chloroformates in gas chromatography as general purpose derivatizing agents. Journal of Chromatography B. 1998;**717**:57-91

[27] Lili L, Xu H, Song D, Cui Y, Hu S, Zhang G. Analysis of volatile aldehyde biomarkers in human blood by derivatization and dispersive liquid–liquid microextraction based on solidification of floating organic droplet method by high performance liquid chromatography. Journal of Chromatography. A. 2010;**1217**: 2365-2370

[28] Dong J-Z, Moldoveanu SC. Gas chromatography–mass spectrometry of carbonyl compounds in cigarette mainstream smoke after derivatization with 2,4-dinitrophenylhydrazine. Journal of Chromatography. A. 2004; **1027**:25-35

[29] Edman P, Begg G. A protein sequenator. European Journal of Biochemistry. 1967;**1**:80-91

[30] Rangarajan M, Darbre A. Identification of amino acid thiohydantoins directly by thin-layer chromatography and indirectly by gas-liquid chromatography after hydrolysis. The Biochemical Journal. 1975;**147**: 435-438

[31] Li M, Biswas S, Nantz MH, Higashi RM, Fu X-A. Preconcentration and analysis of trace volatile carbonyl compounds. Analytical Chemistry. 2012; **84**:1288-1293

[32] Pawliszyn J. Solid Phase Microextraction: Theory and Practice. New-York: Wiley-VCH; 1997

[33] Kataoka H. Automated sample preparation using in-tube solid-phase microextraction and its application–a review. Analytical and Bioanalytical Chemistry. 2002;**373**:31-45

Section 3

Sample Preparation

Chapter 3

Sample Preparation Techniques for Gas Chromatography

Foujan Falaki

Abstract

In gas chromatography (GC), the sample is vaporized and injected onto the head of a chromatographic column. Elution is brought about by the flow of an inert gaseous mobile phase such as helium, argon, nitrogen, carbon dioxide, and hydrogen. In GC, the mobile phase does not interact with molecules of the analyte, and it only transports the analyte through the column. In two general kinds of GC, *gas-solid* chromatography (GSC) and *gas-liquid* chromatography (GLC), the mechanisms of analyte retention in the column are thoroughly different. In GLC, the analyte has been participated between a gaseous mobile phase and a liquid stationary phase. While in GSC, the retention of analytes is the consequence of its physical adsorption onto a solid stationary phase. In comparing of GLC and GSC, more widespread use of GLC has been found in all fields of science. This is mainly due to the semipermanent retention of active or polar molecules and the severe tailing of elution peaks, which is a consequence of the nonlinear character of adsorption process, in GSC. In GC, column efficiency requires that sample be of suitable size and be introduced as a plug of vapor. So, the sample preparation is a very important step in GC. The sample should be injected into a flash vaporizer port located at the head of the column, and its temperature is about 50°C above the boiling point of the least volatile component of the sample. So, the components of the sample should be easily vaporized in this temperature, and they should have high heat resistance not to be decomposed. Both of liquid and solid samples can be introduced to the column. But solid samples are ordinarily introduced as solutions or sealed into thin-walled vials that can be inserted at the head of the column and punctured or crushed from the outside. In order to separate and analyze the gaseous, liquid, and volatile solid samples directly, GC is a suitable analytical equipment. When the analyte sample is nonvolatile, the derivatization and pyrolysis GC techniques are crucial. Gas chromatography can be applied to the solution of many problems in various fields such as drugs and pharmaceuticals, environmental studies like air and clinical samples, petroleum industry, pesticides and their residues, and foods. On the other hand, most samples are not ready for direct introduction into instruments. For organics and volatile organics, the sample preparation procedures can be named as extraction, cleanup, derivatization, transfer to vapor phase, and concentration. So, the basic concept of a sample preparation method is to convert a real matrix into a sample in a format that is suitable for analysis by a separation or other analytical techniques. The goals of sample treatment step are as follows: (1) The capability of using smaller amounts of initial sample, especially for trace analysis. (2) Achieving higher specificity and selectivity in analytical determinations. (3) To improve the potential for automation or online methods and minimize the manual operations. (4) The usage of no or small volumes of organic solvents in order to approach the green chemistry techniques with less wastes and more friendly environment. On the other hand,

IntechOpen

different samples possess a variety of sample treatment methods, for example: (1) In order to treat solid samples and separate a purpose analyte, some enhanced solvent extraction methods include pressurized liquid extraction, microwave- and sonic wave-assisted extraction, supercritical fluid extraction, and superheated water extraction. (2) For analytes in solution, the sample preparation can be attributed to the analyte trapping methods such as -phase extraction, solid-phase microextraction, and stir bar extractions. (3) Also, the extraction of the analytes into a liquid phase can be achieved by other methods like membrane extraction, single-drop microextraction (SDME), and purge and trap. (4) For separation of analytes in the gas phase, trapping analytes from vapor samples and headspace analysis are used. As a result, sample preparation is not only a critical step but also possesses different ways to treat and convert matrix into a suitable sample to inject GC.

Keywords: sample treatment, green chemistry, extraction methods, solid-phase extraction, cleanup

1. Introduction

Most samples are not ready to introduce directly into the column of gas chromatography (GC) instrument [1–7]. So, the sample preparation is the most important step prior to GC determination of an analyte. There might be several processes within sample preparation which depend on the complexity of the sample; the analyte concentration level in the sample and its level need to be analyzed by the GC instrument. On the other hand, sample preparation is often a severe process that accounts for the complexity of the analyte analysis. Instance for organics and volatile organics, the sample preparation procedures can be mentioned such as extraction, cleanup, derivatization, transfer to vapor phase, and concentration.

Before analyzing a sample by GC, the sample preparation procedure should be reviewed to some important constraints such as accuracy, precision, cost, the amount of available laboratories, the analysis time consumption, and the possibility of method automation. Since analytical instruments, like GC, have become quite sophisticated and then provide high levels of accuracy and precision, sample preparation step has been accounted for the majority of the variability. For instance, the sample preparation might involve several discrete steps and also manual handling and take some days, whereas the GC analysis can be performed in a matter of minutes. Therefore, typically two-thirds of the time in GC analysis can be spent on sample preparation. It is worth noting that better improvement in the GC analysis can be brought by significantly simpler sample preparation processes. Some suitable approaches to reducing uncertainty during sample preparation are minimizing the number of steps and using appropriate techniques. On the other hand, the greater the number of steps, the more error there are. So, if it is possible, one or more sample preparation steps should be eliminated. Also, the choice of an appropriate method for sample preparation can improve precision.

The goal should be to choose a combination of sample preparation and GC instrumentation to reduce both of the number of sample preparative steps and relative standard deviation (RSD) and/or increasing precision. Sample preparation step can affect some other quantitative statistical parameters such as limit of detection (LOD), limit of quantitation (LOQ), limit of linearity (LOL), and linear dynamic range (*LDR*). Limit of detection (LOD) is defined as the lowest concentration or weight of analyte which can be determined at a specific confidence level. The lowest concentration level at which a measurement is quantitatively meaningful is called limit of quantitation (LOQ). For all practical purposes, the upper limit of quantitation is the point where the calibration curve becomes nonlinear. This point is called the limit of linearity (LOL). The range of analyte concentration which possesses linearity toward instrumental signal is called linear dynamic range (LDR). Considering all these, the recovery in sample preparation method is an essential parameter which affects quantitative issues such as detection limit, sensitivity, LOQ, and LOL. The sample preparation methods which enhance performance result in larger recovery, higher sensitivity, and lower detection limits. Also, other important parameters in choosing an appropriate sample preparation method include higher speed procedures or use of online methods, low cost, and less reagent consumption or use of greener sample preparation methods.

Before a new sample preparation procedure is used, it must be validated. The different figures of merit should be determined during the validation process. A typical validation process includes the following steps:

1. Determination of the random and systematic errors in terms of precision and bias.

2. Determination of the detection limit for each analyte in the sample.

3. Determination of the accuracy and precision at the concentration range where the GC method is used.

4. Measurement of the linear dynamic range and the calibration sensitivity.

Generally, method validation provides not only a comprehensive picture of merits of a new sample preparation method but also a useful comparison with other existing methods [2].

As mentioned above, the main concept of a sample preparation method is to convert a real matrix into a sample format which is suitable for analysis by a separation or other analytical methods. This can be approached by using a wide range of techniques that have a common list of aims such as [3]:

1. The removal of serious interferences from the sample in order to increase the selectivity of the both separation and detection stages.

2. The increase of analyte concentration and sensitivity.

3. To convert the analyte into a more suitable form to detect, determine, and/or separate.

4. To apply more reproducible techniques which do not depend on the variations of the sample matrix.

Although some traditional sample preparation techniques are still in use, the trends in recent years have been toward to [3]:

1. Using smaller initial sample sizes even for trace analyses.

2. Achieving higher specificity and selectivity.

3. To reduce manual operations and to improve potential for automation or online techniques.

4. To approach to a more environmentally friendly methods (green chemistry) with less or no use of organic solvents and less waste production.

On the other hand, different samples possess a variety of sample treatment methods. Therefore, in this chapter, by paying attention to the type of sample matrix, information required (quantitative or qualitative), and sensitivity required, the sample preparation methods used before GC analyses are discussed.

2. Extraction techniques

The earliest sample preparation method is extraction, in which the analyte of interest is separated from a sample matrix with an optimum yield and selectivity. Two major kinds of extraction include solid-phase extraction (SPE) and liquid-liquid extraction (LLE). In SPE, the analyte can be separated from a solid sample, and in LLE, it is extracted from sample solutions [7]. The solvents, in which the analyte is extracted, may be organic liquids, supercritical fluids, and superheated liquids [3]. Also, the extractor solvent may be bonded to a polymeric support, as in membrane extractions [4]. By optimizing the extraction conditions such as temperature, pressure, and pH of the solution and also appropriately using additives and reagents, the selectivity and yield of the extraction process will be improved.

The basic purpose of all extraction methods is to concentrate the analyte selectively in one phase. Each analyte is distributed between two phases according to the distribution constant, temperature, and relative volumes of the phases. In many of these methods, there is a conflict between the analytes of interest and the other soluble interferences to be quite extracted into the extractor phase. Exhaustive extraction techniques, like Soxhlet extractions, are often designed to provide thorough extractions regardless of the matrix. So, this kind of extraction can be applied to a range of samples such as different soil types but limits selectivity [3].

In order to enhance the selectivity of the extraction process, the supercritical fluid extraction (SFE) was introduced. In comparison with organic solvents, the carbon dioxide solvent is a weaker eluent and more selective extractor medium.

2.1 *Solid-phase* extraction (SPE)

In general, SPE involves four steps:

1. Column preparation or prewash step.

2. Sample loading or the retention of the analytes of interest on the cartridge and/or sorbent.

3. Column postwash to remove undesirable contaminants. In reality, the compounds of interest are retained on the sorbent, while interferences are washed away.

4. Analyte desorption from the cartridge. The adsorbed analytes are recovered by an appropriate eluting solvent.

SPE sorbents are commercially available in three formats:

1. Cartridge.

2. Columns fashioned like syringe barrels.

3. Disks.

Also, sorbent phases can be purchased, and typical column housings are manufactured of polypropylene or glass. In order to contain the column with the sorbent phase, porous frits made of polyethylene, stainless steel, or Teflon can be used [2].

There are some examples for applying SPE as sample preparation step before GC detection of different analytes in a variety of samples. Some of them are pointed below.

Lee et al. reported the determination of endocrine-disrupting phenols, acidic pharmaceuticals, and personal care products in sewage by solid-phase extraction and gas chromatography-mass spectrometry [8]. In this work, an anion exchanger was used as a solid-phase extractor, and a multiresidue method was developed and optimized for the extraction of 21 phenols and acids in sewage. The phenols and acids were then selectively eluted in separate fractions and were converted into volatile derivatives, by suitable reagents, for GC-MS determination.

Stajnbaher et al. studied a multiresidue method for determination of 90 pesticides in fresh fruits and vegetables using solid-phase extraction and gas chromatography-mass spectrometry [9]. In this study, a SPE on a highly cross-linked "*poly(styrene-co-divinylbenzene)*" column was used for cleanup and preconcentration of the pesticides from the water-diluted acetone extracts, and then the pesticides were determined by GC-MS.

2.2 Solid-phase microextraction (SPME)

Miniaturization of analytical processes into microchip platforms designed for micro total analytical systems is a new and rapidly developing field. Solid-phase microextraction (SPME) is a modern technique that consists in direct extraction of the analytes with the use of a small-diameter fused silica fiber coated with an adequate polymeric stationary phase [10]. On the other hand, in two designs of SPME, a thin layer of sorbent is coated on the outer surface of fibers (fiber design), and the inner surface of a capillary tube (in-tube design) is covered. The fiber design can be used as an interface in both GC and HPLC, but in-tube design has been just applied as an easier approaching interface with HPLC. In fiber design, a thin film of liquid polymer or mixture of a solid sorbent with a liquid polymer has covered on the surface of a fused silica core fiber.

The properties of extraction process, by SPME, are as follows [2]:

1. By SPME, samples are analyzed after equilibrium is reached or at a specified time prior to achieving equilibrium.

2. Exhaustive extraction of analyte from the sample matrix is not achieved by SPME.

3. So, SPME operationally encompasses non-exhaustive, equilibrium and preequilibrium, batch, and flow-through microextraction techniques.

4. SPME is directly applicable for field applications in air and water sampling.

5. It does not require continuation of extraction by SPME until equilibrium is reached.

6. A quantitative extraction may be obtained by careful control of time and temperature.

7. SPME is a solventless sample preparation procedure.

8. SPME is compatible with chromatographic analytical systems, and the process is easily automated [11].

9. In conventional SPE, the analyte can be extracted exhaustively (>90%) into the solid phase from a sample medium, while small amount of sample (1–2%) has been introduced into the analytical equipment. But in SPME, although the analyte extraction is non-exhaustive and its small portion has been extracted into the solid phase (about 2–20%), all sample can be injected into the analytical instrument. So, besides high concentration ability and selectivity, SPME possesses another advantage in the ability of using trace analyses [12].

10. SPME facilitates unique investigations, such as extraction from very small samples (i.e., single cells).

11. In SPME, changes in the sample matrix may affect quantitative results (disadvantage).

12. SPME can be used to extract semivolatile organics from environmental waters and biological matrices as long as the sample is relatively clean. Since extraction of semivolatile organics by SPME from dirty matrices is difficult, one strategy for doing it is to heat the sample to drive the compound into the sample headspace for SPME sampling [13].

SPME can be conducted in three modes [2]:

1. Direct extraction, in which the coated fiber is immersed in the aqueous sample.

2. Headspace configuration, for sampling air or the volatiles from the headspace above an aqueous sample. However, headspace techniques are more applicable to volatile organics than to the semivolatile organic compounds.

3. Membrane protection configuration, in which the coated fiber is protected with a membrane, for analyzing the analytes in too much dirty samples.

The SPME procedure is performed through two separate steps:

a. At first, the solid sorbent is immersed into the sample medium for a specific period of time. This step is used for both of fiber and in-tube designs.

b. Then, the solid sorbent, either fiber or in-tube design, is interfaced with GC and HPLC (or capillary electrophoresis) instruments for thermal and solvent desorption processes, respectively.

As discussed before, SPME has been introduced as a solventless extraction method, in which a fused silica fiber has been coated with a *thin* film of sorbent, to separate volatile analytes of interest from a matrix sample. Usually, the fiber is placed into a syringe needle which is protected for easy penetration into the sample and GC vial septa. Analyte extraction and analysis depend on the fiber type and its thermal desorption into a GC inlet. There are two approaches to SPME sampling of volatile organics:

1. Direct sampling. In this approach, the fiber is placed directly into the sample matrix.

2. Headspace sampling. In this approach, the fiber is placed in the headspace of the sample.

Choosing between direct immersion and headspace SPME is relatively straightforward. Direct immersion SPME is warranted for liquid and solution samples which are used in solid-phase and liquid-liquid extraction methods. Headspace SPME is considered for extraction of volatile specious, with normal boiling point less than 200°C, from solid and liquid samples. For higher boiling point analytes, direct immersion SPME is probably necessary. Headspace is more preferred for especially complex or dirty samples due to fouling the fiber coating in a direct immersion process.

SPME fibers have different coatings, and there is no single coating for extraction and separation of all volatile organics from a sample. Therefore, different types of coatings with different polarities are applied on SPME fibers. Currently, three types of fiber coatings are commercially available: (1) nonpolar, (2) semipolar, and (3) polar coatings. There are several SPME fiber coatings commercially available. These range in polarity from polydimethylsiloxane (PDMS), which is nonpolar, to "*carbowax-divinylbenzene*" (CW-DVB), which is highly polar. The nonpolar fibers are more commonly used for headspace SPME as the majority of volatile analytes tend to be nonpolar or slightly polar. The advantage of using different fiber polarities is that using a matched polarity fiber, as polar coated for a polar analyte, makes extraction selectivity be enhanced. On the other hand, there is less of a chance of extracting interfering compounds along with the analyte of interest, and an organic matrix is not a problem.

Fiber coating thickness is a second parameter that should be considered to select a fiber for both direct immersion and headspace SPME. The PDMS coating is commercially in hand in three thicknesses: 100, 30, and 7 μm. The 100-μm-thick fiber is generally applied for highly volatile compounds or when a larger organic matrix volume is used. The 7-μm-thick fiber is used for less volatile compounds.

Once the fiber is chosen, extraction conditions must be optimized. There are many variables such as (1) extraction time, (2) sample volume, (3) agitation, (4) temperature, and (5) sample matrix.

1. As extraction time is increased, a plateau in peak area is reached. So, this represents the time required for the system to reach the equilibrium and is the optimized extraction time. Most headspace SPME methods are completed in less than 5 min, while direct immersion SPME may require more than 30 min. Also, direct immersion SPME is highly matrix dependent.

2. The sensitivity of a SPME method is proportional to the number of moles of analyte recovered from the sample. As the sample volume increases, analyte recovery increases too. But in very dilute samples, larger sample volume results in slower kinetics and higher analyte recovery.

3. In many extraction methods, the agitation method affects both the extraction time and efficiency. In direct immersion SPME, agitation is often accomplished with a magnet and a stirrer. So, the stirring rate should be optimized and constant during the extraction process. Also, the fiber should be off-centered in the vial so that liquid is moving quickly around it. Agitation can also be achieved by physical movement of the fiber or by movement of the sample vial and/or sonication.

4. Extraction temperature can also be an important factor, especially in headspace SPME analyses. Despite of GC headspace analysis, increasing the temperature in SPME makes the extraction sensitivity decrease.

5. By modifying the sample matrix, the extraction recovery can be improved. There are two ways to modify the sample matrix: (a) adjusting the sample pH

or its salt content and (b) dissolving the solid sample in a proper solvent like water or a strongly aqueous solution. In similarity with classical liquid-liquid extractions, modifying the pH can change the extraction behavior.

Also, the SPME-GC injection system must be optimized. When the SPME interfacing GC is used, the GC injection system is typically done under splitless conditions. Since there is not any solvent and accommodation of the sample solvent, there is no need of specific small internal diameter glass liners, which are often used [2]. The main consideration is to transfer the analytes in the shortest possible time out of the fiber coating and in focusing the analytes into the sharpest bands possible. For semivolatile compounds, inlet optimization is very simple, and classical splitless inlet conditions can be used. A typical condition would be a temperature of about 250°C; a sufficient head pressure can maintain optimum GC column flow and an initial column temperature at least 100°C below the normal boiling point of the analyte. For volatile analytes, the optimization of the inlet is more difficult. So, keeping the initial column temperature at enough low level to refocus these analytes is often not possible, without cryogenics. The inlet must therefore be optimized to provide the fastest possible desorption and transfer to the GC column, while the GC column is maintained as cool as possible to achieve any focusing that is possible.

There are some examples for applying SPME as sample preparation step before GC detection of different analytes in a variety of samples. Some of them are pointed below.

Goncalves et al. studied solid-phase microextraction-gas chromatography-(tandem) mass spectrometry as a tool for pesticide residue analysis in water samples at high sensitivity and selectivity with confirmation capabilities. In this study, for SPME extraction a *"poly(dimethylsiloxane)-poly(divinylbenzene)"*-coated fiber was selected [14].

Yonamine et al. studied solid-phase microextraction-gas chromatography-mass spectrometry and headspace-gas chromatography of tetrahydrocannabinol (THC), amphetamine, methamphetamine, cocaine, and ethanol in saliva samples. In this study, at first saliva samples were submitted to an initial headspace procedure for ethanol determination by a GC-flame ionization detector. Then, two consecutive fiber solid-phase microextractions were carried out: THC was extracted by submersing a polymeric fiber, and amphetamine, methamphetamine, and cocaine were subsequently extracted after alkalinization [15].

2.3 Molecularly imprinted polymer (MIP) adsorbent in SPE and SPME

Molecularly imprinted polymer (MIP) is an alternative kind of sorbent which can be applied for *solid-phase extractions* and solid-phase microextractions. MIP is a polymeric sorbent which is produced in the presence of a target analyte, as a molecular template. Once the template is washed and removed through the polymer, some selective recognition sites has been remained in the polymeric sorbent for selective extraction of the analyte target. By using MIP as the sorbent, the surface contact area between the sorbent and the sample is much greater than in the coated fiber or coated inner surface tubing SPME procedures described earlier [2]. MIP inherent advantages include reusability, simplicity, low cost, high affinity and selectivity for target molecule, and physical and chemical stability over a wide range of experimental conditions and solvents [16].

Some of those applications are discussed below.

Djozan et al. studied preparation and evaluation of solid-phase microextraction fibers based on monolithic molecularly imprinted polymers for selective extraction of diacetylmorphine and analogous compounds. The main purpose of this research was to develop a technique for fabrication of a monolithic and robust solid-phase microextraction on the basis of MIP interfacing with GC and GC-MS analysis for selective extraction and structural analysis of diacetylmorphine, respectively. On the other hand, a very simple approach has been developed for the fabrication of SPME fiber from diacetylmorphine-imprinted polymers which were subsequently used for extraction of diacetylmorphine and then analyzed with GC and GC-MS [16].

Rehim studied new trend in sample preparation [17]: online microextraction in packed syringe for liquid and gas chromatography applications and determination of local anesthetics in human plasma samples using gas chromatography-mass spectrometry. In this study, local anesthetics in plasma samples were used as model substances, and the method was developed and validated for microextraction in packed syringe (MEPS) online with GC-MS. MEPS and SPE procedures have some differences. In MEPS method, the sorbent packing is placed directly into the syringe, not into a separate column, as it is done in SPE. So, a separate robot does not need for applying the sample into the separation phase, as it is done in SPE. Also, the packed syringe can be applied several times for different samples [17].

2.4 Stir bar sorptive extraction (SBSE)

Stir bar sorptive extraction (SBSE) is used for the extraction of trace amounts of organics from aqueous food, environmental, and biological samples. A stir bar has been covered with a sorbent phase and placed into the sample solution to separate the analyte of interest. Although SBSE procedures are not exhaustive, more quantitative extractions can be achieved than those of SPME procedures.

The coated stir bar is usually used to stir the sample solution for a specialized period of time, depending on the sample volume and stirring speed, until approaching equilibrium.

SBSE improves on the low concentration capability of in-sample solid-phase microextraction (IS-SPME). Also, SBSE can be applied to headspace sorptive extraction (HSSE) [2].

Some of SBSE applications with GC analysis are discussed below.

Nakamura et al. studied simultaneous determination of alkylphenols and bisphenol A in river water by stir bar sorptive extraction with in situ acetylation and thermal desorption-gas chromatography-mass spectrometry. In this study, SBSE was used for the sample enrichment of seven alkylphenols and bisphenol A in river water. Also, in situ derivatization in aqueous samples was performed with acetic acid anhydride as acetylation reagent [18].

The extraction phase on the stir bar in SBSE is critical for the performance of both extraction and thermal desorption. The sol-gel coating technology possesses the potential to prepare thermally stable coatings [19].

Guan et al. studied determination of organophosphorus pesticides in cucumber and potato by stir bar sorptive extraction. In this study, organophosphorus pesticides (OPPs) in vegetables were determined by SBSE and capillary GC with thermionic specific detection (GC-TSD). Hydroxy-terminated polydimethylsioxane (PDMS) prepared by sol-gel method was used as extraction phase [19].

2.5 Soxhlet extraction

Soxhlet extraction was accepted as a standard method for the extraction of semivolatile and nonvolatile organics by the US Environmental Protection Agency (EPA 3540C0) and also the extraction of fat in cacao products by the Association of Official Analytical Chemists (AOAC 963.15). Soxhlet extraction was *introduced by Franz Ritter von Soxhlet in 1879*. It had been the most extensive applied technique

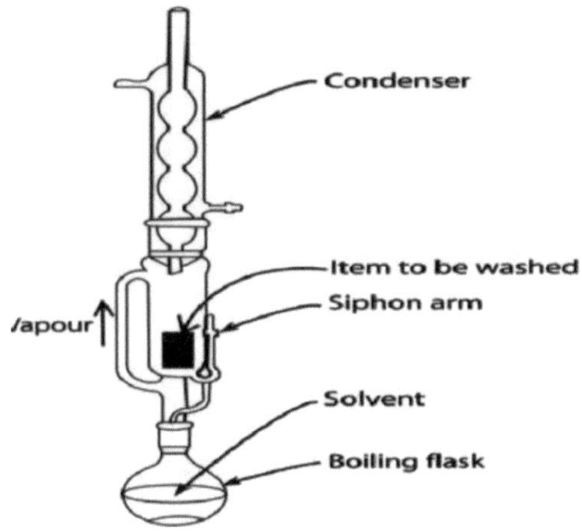

Figure 1.
The schematic diagram of Soxhlet apparatus.

till the other modern extraction methods were developed in the 1980s. Nowadays, Soxhlet is still applied for the extraction of semivolatile organic compounds from solid samples. Soxhlet extraction is a classical method which is operated under atmospheric pressure, in high temperature or under ultrasonic irradiation. In this technique, relatively large volumes of organic solvents are usually used, and it is a time-consuming technique [2].

Soxhlet apparatus has three components, and its schematic diagram is shown in **Figure 1** [2]:

1. The top part is a solvent vapor reflux condenser.

2. The middle part is a thimble holder with a siphon device and a side tube.

3. The bottom part is a round-bottomed flask which connects to the thimble holder.

4. A porous cellulous sample thimble is filled with sample solution and inserted into the sample thimble holder. Usually, 300 ml of solvents is introduced to flask for 10 g of a sample. The flask is heated slowly on a heating mantle, and the solvent vapor goes toward the reflux condenser and, after condensing, drips back to the thimble chamber. When the analyte reaches the top of the sample thimble holder, it is transferred back into the bottom flask via a siphon device. This cycle is repeated many times for a predetermined period of time. Since the boiling points of analytes are usually higher than those of solvents, the analytes accumulate in the flask and the solvents recirculate. Finally in each cycle, the analyte can be extracted with fresh solvents.

The properties of Soxhlet extraction are as follows [2]:

1. In Soxhlet extraction, the extraction is slow and can take between 6 and 48 h. On the other hand, it is a time-consuming technique (its drawback). It is mainly due to the analyte that is extracted with cooled condensed solvent.

2. The extract volume is relatively large (its drawback). So, the evaporation step is usually needed to concentrate the analytes before the analysis.

3. The sample size is often *10 g* or more, and multiple samples can be extracted on separate Soxhlet units.

4. Soxhlet is a rugged and well-established technique.

5. Relatively large solvent consumption (its drawback).

An automated Soxhlet extraction (Soxtec) was approved by the EPA (EPA 3541) in 1994 for the extraction of semivolatile and nonvolatile organic compounds [2]. Automated Soxhlet extraction is relatively faster than Soxhlet extraction, with lower consuming organic solvents [2]. In this method, the extraction is performed in three stages:

- In the first stage, a thimble containing the sample is immersed in the boiling solvent for about 60 min. Since the contact between the solvent and the sample is more vigorous and the mass transfer in a high-temperature boiling solvent is more rapid, extraction here is faster than in Soxhlet.

- In the second stage, the sample thimble is placed above the boiling solvent. Then, the condensed solvent drips into the sample and extracts the organics and falls back into the solvent reservoir as well. This stage is similar to traditional Soxhlet and takes usually 60 min.

- In the third stage, the solvent is evaporated, and a concentration step happens for 10–20 min.

Li et al. studied the determination of organochlorine pesticide residue in ginseng root by orthogonal array design Soxhlet extraction and gas chromatography. In this study, a method involving four-factor-three-level orthogonal array design was developed. The orthogonal array designs included extracting solvent component, particle size, solvent overflow recycle, and time needed for the optimization of extracting nine organochlorine pesticides from ginseng root, followed by capillary GC-electron capture detector and MS detector [20].

2.6 Ultrasonic extraction

Ultrasonic extraction, also known as sonication, uses ultrasonic vibration to ensure intimate contact between the sample and the solvent. Sonication is relatively fast, but the extraction efficiency is not as high as some other techniques. Also, ultrasonic irradiation may decompose some of organophosphorus compounds.

Before the sonication is used for real sample, the selected solvent system and optimum conditions for adequate extraction of the target analytes from reference samples should be investigated.

A typical sonication device can be equipped with a titanium tip. The sample is usually dried with anhydrous sodium sulfate and mixed with a certain volume of selected solvent. The disruptor horn tip is positioned just below the surface of the solvent, yet above the sample. Extraction can be carried out in duration as short as 3 min. After extraction, the extract is filtered or centrifuged, and also some form of cleanup is needed before analysis [2].

The ultrasonic extraction (USE) is a very versatile technique due to the possibility of selecting the solvent type or solvent mixture that allows the maximum extraction efficiency and selectivity. In USE, several extractions can be done simultaneously, and no specialized laboratory equipment is required (advantage). But it is not easily automated (disadvantage) [21].

Goncalves et al. studied the assessment of pesticide contamination in soil samples from an intensive horticulture area, using ultrasonic extraction and gas chromatography-mass spectrometry. In this study, the application of an USE method combined with GC and GC-MS for the analysis of some pesticides in soil samples was investigated. The USE technique was used to separate the pesticides from the soil samples [21].

2.7 Supercritical fluid extraction (SFE)

In supercritical fluid extraction (SFE), supercritical fluids possess specific properties which make them facilitate the extraction of organics from solid samples. Two configuration of SFE operations are on- or off-line mode. In the online operation, SFE is matched directly to an analytical instrument like GC, supercritical fluid chromatography (SFC), and HPLC. Off-line SFE, as its name implies, is a stand-alone extraction method independent of the analytical method to be applied. Off-line SFE is more flexible and easier to perform than that of the online procedure. It allows the Extract to be available for analysis by different techniques [2].

A supercritical fluid (SF) is a substance above its critical temperature and pressure. Also, it is an interface between gas and liquid. In fact it is not a liquid and or a gas, it is a SF.

CO_2 has a low supercritical temperature (31°C) and pressure (73 atm). It is nontoxic and nonflammable and also is available at high purity. So, carbon dioxide has become the solvent of interest for most SFE applications. Supercritical CO_2 is nonpolar and without permanent dipole moment; therefore, it can be utilized to extract nonpolar and moderately polar compounds from matrices. For the extraction of polar compounds, supercritical N_2O and $CHClF_2$ are more efficient. But these SFs are not environmentally friendly and they are not used in routine analysis [2].

SFE has gained increased attention as a good candidate instead of conventional liquid solvent extraction. This is mainly due to significant properties of supercritical fluids (SFs) such as their high diffusivity and low viscosity which make them extract selectively different chemicals without additional cleanup steps and so use little sample amounts [22].

Rissato et al. studied the supercritical fluid extraction for pesticide multiresidue analysis in honey and determination by gas chromatography with electron-capture and mass spectrometry detection. In this study, SFE procedure was used to separate some pesticides from honey samples, and it was compared with liquid-liquid extraction method [22].

2.8 Accelerated solvent extraction (ASE)

The other names of accelerated solvent extraction (ASE) are pressurized fluid extraction (PFE) and pressurized liquid extraction (PLE). Conventional solvents are used in ASE at high temperature (100–180°C) and pressure (1500–2000 psi) to increase the extraction percentage of organic compounds from solid samples.

Supercritical fluid extraction is matrix dependent and usually needs the addition of organic modifiers. ASE was developed to overcome these limitations. Although it was expected that conventional solvents would be less efficient than supercritical fluids, the results turned out to be quite the opposite. In many cases, extraction was faster and more complete with organic solvents at elevated temperature and pressure than with SFE [2].

The elevated pressure and temperature used in ASE affect the solvent and sample properties and their interactions as well. ASE properties include the following [2]:

1. Under higher pressure, the extraction would be performed at higher temperature values. This is mainly due to the increase of the solvent boiling point.

2. At higher pressures, the solvent penetration into the sample medium would be increased, and so the extraction of the interested analyte may be facilitated from the matrix.

3. At higher temperatures, the mass transfer and solubility of the analyte are enhanced.

4. The elevated temperature can reduce the power of analyte-sample bonds like dipole, hydrogen, and van der Waals interactions.

5. High temperature decreases the solvent viscosity and surface tension and so enhances solvent penetration into the matrix medium.

6. Therefore, faster extractions and better analyte recoveries can be achieved by ASE procedures.

ASE process has some steps mentioned below:

1. The extraction cell is filled with the sample medium.

2. Then, the solvent is entered in.

3. And, the cell temperature and pressure are increased to the desired level. The necessary time to enhance the temperature can be between 5 and 9 min (for up to 200°C).

The above steps are referred to the prefill method. If before addition of solvent the sample is warmed, the process is mentioned as preheat method. In comparison of the two procedures with each other, the prefill method is usually preferred [2].

A. Pastor et al. studied the determination of PAHs in airborne particles by accelerated solvent extraction and large-volume injection-gas chromatography-mass spectrometry. The procedure included extraction of some PAHs by accelerated solvent extraction (ASE) followed by gel permeation chromatography (GPC) cleanup and GC-MS detection of PAHs. In this study, the hexane-acetone mixture (1:1 v/v) gave the best recoveries when ASE parameters were fixed at 125°C and 1500 psi and a total time of 10 min [23].

2.9 Microwave-assisted extraction

It should be noted that microwave-assisted extraction (MAE) is different from microwave-assisted acid digestion. The former uses organic solvents to extract organic compounds from solids, while the latter uses acids to dissolve the sample for elemental analysis with the organic contents being destroyed. MAE is applied for the extraction of semivolatile and nonvolatile compounds from solid samples.

In general, organic extraction and acid digestion use different types of microwave apparatuses, as these two processes require different reagents and experimental conditions. The basic components of a microwave system include a microwave generator (magnetron), a waveguide for transmission, a resonant cavity, and a power supply. There are two types of laboratory microwave units:

1. Closed extraction vessels under elevated pressure.

2. Open vessels under atmospheric pressure.

In the liquid and solid states, molecules do not rotate freely in the microwave field, despite of gaseous molecules; therefore, no microwave spectra can be observed. Liquid- and solid-state molecules respond to the radiation differently, and this is where microwave heating comes in. During microwave heating procedure, electromagnetic energy would be changed to heat. This is mainly due to the ionic conduction and dipole rotation of the molecules which are imposed. Ionic conduction is concluded from the ion mobility in a solution under an electromagnetic field, and then, the heat is produced. Dipole rotation means that the directions of dipole rotations are changed under microwave irradiation. When a polarized molecule is imposed in an electromagnetic field, it can rotate around its axis at a rate of 4.9×10^9 times per second. So, with the larger molecular dipole moments, the more vigorous oscillations of molecules are obtained under a microwave field.

The proper choice of solvent is the key to successful extraction in MAE. In general, three types of solvent system can be used in MAE:

1. Solvent(s) of high dielectric coefficient.

2. A mixture of solvents of high and low dielectric coefficient.

3. A microwave transparent solvent used with a sample of high dielectric coefficient.

Zhou et al. studied the microwave-assisted extraction followed by gas chromatography-mass spectrometry for the determination of endocrine-disrupting chemicals in river sediments. In this study, the most efficient extraction (>74%) of the analyte was achieved by choosing methanol as the solvent, 110°C and 15 min, as the extraction temperature and time, respectively. The cleanup step was performed by passing the extracts through a non-deactivated silica gel column [24].

2.10 Headspace extraction

From an analytical point of view, volatile organic compounds (VOCs) are organic materials whose vapor pressures are greater than or equal to 0.1 mmHg at 20°C. Many VOCs are environmental pollutants, and in most cases of their analyses, the analytes are transferred to a gas-vapor phase and then analyzed by GC techniques [2].

Generally, the analysis of pure volatile compounds is simple, and the volatile analyte can be injected directly into a GC column [25]. However, the challenge is to extract the analytes from the matrix samples such as soil, food, cosmetics, polymers, and pharmaceutical raw materials. Headspace extractions are approaches to this and are divided into two categories: static headspace extraction (SHE) and dynamic headspace extraction (purge and trap) [2].

Static headspace extraction is known as equilibrium headspace extraction or simply as headspace. This technique has been available more than 30 years, so its instrumentation is both mature and reliable. In this technique, the extraction method includes the following [2]:

1. A sample, either solid or liquid, is put in a headspace autosampler (HSAS) or vial.

2. The sample vial is brought to a constant temperature and pressure, and the volatile analytes diffuse into the headspace vessel.

3. When the analyte concentration in the headspace part of the vessel reaches to an equilibrium level with respect to its concentration in the sample, the vial is connected to the GC column head, and then, a portion of the headspace is introduced into a GC for detection. This analyte transfer is due to a pressure drop between the vessel and the GC inlet pressure.

4. The vial is again isolated. For automated systems, this sampling procedure can be repeated by the same or the next vial.

The advantage of static headspace extraction is the ease of initial sample preparation. Usually for qualitative analysis, the sample can be placed directly into the headspace vial and analyzed with no additional preparation procedures. But for quantitative analysis, it may be vital to know the optimized matrix effects to gain good sensitivity and accuracy.

For large solid samples, it may be needed to change the physical state of the sample matrix. Two approaches in differentiating the sample state are to powder the solid sample and to disperse it into a liquid.

By crushing the solid sample, the surface area available for the volatile solute to distribute into the headspace phase is enhanced. So, the solute is distributing between a solid and the headspace phases. But in the second procedure, dispersing the solid into a liquid is preferred because the analyte partitioning process into the headspace often reaches the equilibrium faster. Therefore, by choosing a suitable solvent with high affinity toward the volatile analytes, the problems with sample and standard transfer from volumetric flask to headspace vials can be eliminated [2]. Some experimental factors affecting SHE should be optimized to improve extraction efficiency, sensitivity, quantitation, and reproducibility. These experimental variables include vial and sample volume, temperature, pressure, and the form of the matrix itself.

For the analysis of trace amounts of analytes, or where an exhaustive extraction of the analyte is required, purge and trap or dynamic headspace extraction (DHE) is more preferred than SHE. This technique is used for both solid and liquid samples. The samples can be biological, environmental, industrial, pharmaceutical, and agricultural. In DHE, there is no equilibrium between its concentration in the gas and matrix phases. Instead, they are removed continuously from the sample by a gas flow. This provides a concentration gradient between two mentioned phases which makes the exhaustive extraction of the volatile analytes.

A typical purge and trap system consists of the following:

1. A purge vessel.

2. A sorbent trap.

3. A six-port valve.

4. Transfer lines.

A purge and trap cycle consists of several steps: (1) purge, (2) dry purge, (3) desorb preheat, (4) desorb, and (5) trap bake. Each step is synchronized with the operation of the six-port valve and the GC [or GC-MS (mass spectrometer)]. The mentioned steps in a purge and trap cycle can be explained as follows:

1. An aqueous sample is introduced into the purge vessel.

2. The valve is set to the purge position. A purge gas (typically, helium) breaks through the sample continuously and sweeps the volatile organics to the trap, where they are retained by the sorbents. Then, the gas is vented to the atmosphere.

3. The purging step consists of purge, dry purge, and preheating. However, the purge step takes about 10–15 min, and the flow rate of helium is about 40 ml/min. After purging, while the trap is at the ambient temperature, the purge gas is transferred directly into the trap without passing through the sample. This step is called dry purge. The main objective of this step is to remove the water which has been accumulated on the trap. Dry purging often takes place between 1 and 2 min. Then, the purge gas is turned off, and the trap is heated to about 5–10°C below the desorption temperature. Preheat makes the subsequent desorption faster.

4. When the purging step is complete, the trap is heated, from 180 to 250°C, to desorb the analytes into the GC column to be analyzed. On the other hand, it is back-flushed with the GC carrier gas. So, the preheat temperature is reached, and the six-port valve is rotated to the desorb position to initiate the desorption step. Desorption time is about 1–4 min and depends on the carrier flow rate in GC instrument. For instance, the trap desorption time is short at the high flow rate, and so, a narrowband injection is achieved. The flow rate of the desorb gas should be selected in accordance with the type of GC column used. On the other hand, the operational conditions of the purge and trap must be compatible with configuration of GC system. With a packed GC column, higher carrier gas (desorb gas) flow rates can be applied. Usually, the optimum flow rate is about 50 ml/min. Capillary columns require lower flow rate and are often preferred over the packed one for better resolution.

5. In the trap baking step, after the desorption step, the valve is readjusted in the purge position. The trap condition is adjusted at desorption temperature, or 15°C upper than it, for 7–10 min. The objective of this step is to remove possible contaminants and eliminate sample transport.

6. After the trap baking step, the trap temperature is diminished and the next sample can be extracted. In each step, the conditional parameters such as temperature, time, and flow rate should be the same for all of the samples and calibration standards.

The trap is usually a stainless steel tube 3 mm in inside diameter (ID) and 25 mm long packed with multiple layers of adsorbents, and it should do the following steps:

1. Retain the analytes of interest, but do not introduce impurities.

2. Allow rapid injection of analytes into the GC column.

The sorbents are often arranged in layers to increase the trapping capacity. During purging process, the purge gas reaches the weaker sorbent at first, and only less volatile organics are retained. But more volatile compounds just pass through this layer and then are trapped by the other stronger adsorbent layers. During desorption process, the trap is heated and back-flushed with the GC carrier gas. However, the less volatile compounds have never been in contrast with the stronger adsorbents, and so, the reversible adsorptions can be achieved.

To trap volatile organic compounds, the substances such as Tenax, silica gel, activated charcoal, graphitized carbon black (GCB or Carbopack), carbon molecular sieves (Carbosieve), and Vocarb are usually used [2]. Tenax is not only a porous but also a hydrophobic polymer resin based on 2,6-diphenylene oxide, with low affinity for water. So, highly volatile and polar compounds are seldom adsorbed on Tenax. Tenax should not be heated to temperatures upper than 200°C, because of its decomposition under high temperatures. The two types of Tenax are Tenax TA and Tenax GC. The former has higher purity and is more preferred for trace analysis. Silica gel is hydrophilic and is an excellent candidate for trapping polar compounds. Also, it is a stronger sorbent than Tenax. The problem is that water can be retained on the gel. Charcoal, as another stronger sorbent than Tenax, is hydrophobic and is mainly used to trap very volatile compounds such as dichlorodifluoromethane, a.k.a. Freon 12. These compounds can break through Tenax and silica gel. Conventional traps like Tenax, silica gel, and charcoal are usually used in series. If the boiling points of the analytes are above 35°C, Tenax itself will be suitable, and so, silica gel and charcoal can be ignored. Graphitized carbon black (GCB), as an alternative sorbent to charcoal and silica gel, has both the hydrophobic property and the trapping capacity similar to Tenax. Also, it is often used along with carbomolecular sieves and can trap highly volatile compounds. Vocarb is a highly hydrophobic activated carbon which can diminish water trapping and be purged fast. Vocarb is usually operated with an ion-trap mass spectrometer, which can be affected by trace levels of water or methanol. GCB, carbon molecular sieves, and Vocarb possess high thermal stability and can be operated at higher desorption temperatures than those that Tenax can be done [2].

The transfer line between the trap and the GC column is often made of nickel, deactivated fused silica, and silica-lined stainless steel tubing. By using these inert materials, the active sites which can interact with the analytes are eliminated. On the other hand, the transfer line is kept at a temperature higher than 100°C to avoid the condensation of water and the volatile organics. Also, the six-port valve which controls the gas flow path is also heated above 100°C to avoid condensation.

2.10.1 Interfacing purge and trap

As noted above, the operational conditions of purge and trap must be adaptable with the GC system configuration. For example, megabore capillary columns (0.53 mm ID or larger) are typically used at a flow rate of 8–15 ml/min. Since desorption process is slower at such flow rates, the column is usually cooled to 10°C or less temperatures at the stating of the GC run to retain the very volatile compounds. Sub-ambient cooling may be eliminated by using a long column (60–105 m) with a thick film stationary phase (3–5 μm). However, this flow rate is still too high for a GC-MS analyzer. So, a GC-MS interface like a jet separator or an open-split interface should be applied to decrease the carrier flow rate in the mass detector. However, an open-split interface makes a reduction in the analytical sensitivity due to entering just a portion of analyte into the detector [2].

Narrow-bore capillary columns (0.32 mm ID or smaller) with MS detector are commonly operated at lower flow rates (less than 5 ml/min). There are two ways to couple purge and trap with this type of columns:

1. To desorb the trap at a high flow rate and, then, with a split injector, split the flow into the GC instrument. So, a fast injection is obtained without significant loosing of the analytical sensitivity.

2. To desorb or refocus the analytes on a second trap and use a low desorb flow rate. At this flow rate, the time of desorption is too long to achieve a narrow bandwidth injection. A cryogenic trap is often used as a second trap and made of a short piece of uncoated fused silica capillary tube. It is cooled to −150°C by liquid nitrogen to refocus the analytes. After refocusing the analytes, the cryogenic trap is heated quickly to 250°C to desorb the analytes into the GC column.

A moisture control device is another interface which must be used. The purge gas, which is coming from purge vessel, is saturated with water, and so water can be collected on the trap and later released into the GC column during trap heating. Water decreases column efficiency and interferes with some certain detectors such as PID and MS. The column can also be plugged by ice if cryogenic trapping is used. Therefore, water requires to be removed before entering the GC. Two water management techniques are ordinarily applied [2]:

1. To have a dry purge step prior to the desorption process. However, by this approach, some hydrophilic sorbents such as silica gel are not compatible.

2. To use a condenser between the trap and the GC instrument. The condenser is made of inert materials such as a piece of nickel tube. During desorption, the condenser is maintained at ambient temperature, and water is condensed and removed from the carrier gas. After completing the desorption process, the condenser is heated and the water is vented.

Lacorte et al. studied an automated technique based on purge and trap coupled to gas chromatography with mass spectrometric detection for the trace determination of five of the most important water odorants. Analytes were purged from 20 ml of water sample containing sodium chloride at ambient temperature and trapped on a Tenax sorbent by a flow of He. The desorption step was done with helium, purge gas, and temperature programming. The desorbed analytes were directly transferred to a gas chromatograph with a mass spectrometer detector for separation and determination [26].

2.11 Membrane extraction

Among a wide variety of separation methods, membrane extraction and/or transport of analytes through the membranes is a powerful technique for their concentration, separation, and recovery. In this method, the sorption and desorption steps are combined into a one-step process, and, because of its simplicity, low cost, and high efficiency, it possesses an important role in biology, chemistry, and separation sciences; therefore, the efforts for developing of these types of sample treatment methods are increased [5]. In membrane transport, the sample is in contact with one side of the membrane, which is referred to the feed (or donor) phase. Also, the membrane phase serves as a selective barrier. The analytes pass through the membrane phase toward its other side, which is referred to the permeation (or acceptor) phase. Sometimes, the permeated analytes are swept by another phase like either a gas or a liquid. Its schematic diagram is shown in **Figure 2**.

A membrane can be accompanied with an instrumental analysis for online analysis (its advantage). Specially, a mass spectrometer or gas chromatograph can be applied as the detector device. Once a membrane is coupled to the mass spectrometry (MIMS), the membrane can be put in the vacuum compartment of the mass spectrometer. The permeated analytes are directly introduced into the

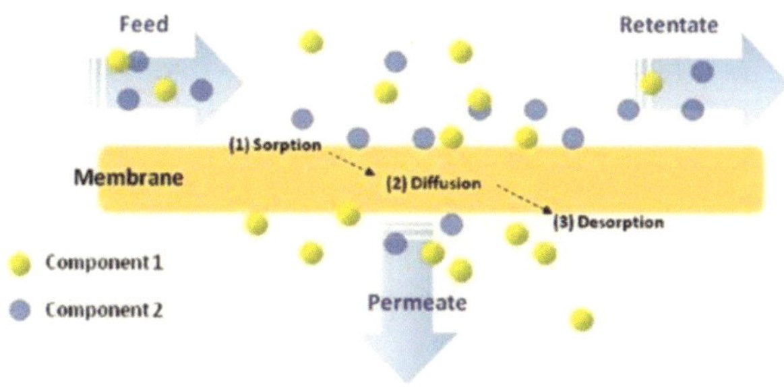

Figure 2.
The schematic diagram of membrane extraction.

ionization chamber of the MS instrument. In membrane introduction gas chromatography, a sorbent trap is interfaced between the membrane and the GC. Then, the permeated analytes are carried by a gas stream to the trap for preconcentration step. After completing the trap or preconcentration step, the trap is quickly heated to desorb the analytes into the GC column, as a narrowband injection. For instance in a GC connection, an aqueous sample from the loop of a multiport injection valve is injected into the hollow fiber membrane module by a N_2 stream. The gas pushes the sample through the membrane fibers, and so the organic analytes permeate to the acceptor phase. Then, they are swept to a micro-sorbent trap by a countercurrent nitrogen stream. After completing the extraction of the analytes on the trap, during a predetermined period of time, the trap is electrically heated to desorb the analytes into the GC column [27].

For matrix samples, GC has gained a good potential of choice, because of its excellent separation ability. Tandem MS has been introduced as a faster alternative technique to GC separation, but such these instruments make higher costs. In membrane-based methods, limit of detections are especially in the parts per thousand (ppt) to parts per billion (ppb) range.

The main drawback in membrane extraction coupled with a GC instrument is the slow permeation through the polymeric membrane and the aqueous boundary layer. This problem is much less than it in membrane introduction mass spectrometer (MIMS). The reason is that the vacuum in the mass spectrometer makes a high partial pressure gradient for mass transfer.

The time needed to complete the permeation process is mentioned as lag time. Another disadvantage of membrane extraction is that the lag and/or transport time can be significantly longer than the time of sample residence in the membrane phase. This is mainly due to the boundary layer effects. When the carrier fluid is an aqueous stream, a static boundary layer is formed between the membrane and the aqueous phase. Since the analytes are being stuck in the boundary layer, the gradient for mass transfer decreases and the transport time enhances. Sample dispersion is another cause of the long lag time in flow injection techniques where an aqueous carrier fluid is used. Axial mixing of the sample with the carrier stream causes dispersion. So, the sample volume increases, and longer residence time in the membrane phase is obtained. Dilution reduces the concentration gradient across the membrane, which is the driving force for diffusion [2, 5].

Membrane pervaporation (permselective "evaporation" of liquid molecules) is the term used to describe the extraction of volatile organics from an aqueous matrix to a gas phase through a semipermeable membrane. The extraction of volatiles from a gas sample to a gaseous acceptor across the membrane is called permeation, which is the mechanism of extraction from the headspace of an aqueous or solid sample. In pervaporation process, the organic analytes of interest move from the bulk aqueous sample solution into the membrane phase and dissolve into it. Then, the analytes diffuse across the membrane phase and permeate into the acceptor or permeate phase and evaporate into the gas phase, as well. An additional step is occurred in headspace sampling mode, and the analytes transport into the headspace phase from the bulk aqueous phase. In both cases, the concentration gradient across the membrane is the driving force for the analyte transport across the membrane. Its schematic diagram is shown in **Figure 3**.

2.11.1 Membrane modules

Membranes can be categorized both based on its structure in two kinds, porous and nonporous, and based on its geometry in two types, flat sheet and hollow fiber. Membranes which are applied in pervaporation and gas permeation are especially hydrophobic and nonporous silicone (polydimethylsiloxane or PDMS) membranes. Aqueous organics dissolve into the membrane phase and are extracted, while the aqueous contaminants are unextracted into the membrane. The microporous membranes in pervaporation are usually made of polypropylene, cellulose, or Teflon. The disadvantage of this membrane is to permit the passage of large quantities of water. Usually, water must be removed before it enters the analysis instrument.

As understood the name, flat-sheet membranes are flat, like a sheet of paper, and can be made as thick as less than 1 mm. However, the typical holders are necessary to hold them in place. In-tube hollow fiber membranes are 200–500 mm in diameter and also allow fluids to flow both inside and outside. Hollow fibers are self-supported and offer the advantage of larger surface area per unit volume and high packing density. A large number of parallel fibers can be packed into a small volume.

2.11.2 Optimization of membrane extraction

Several factors, which affect the extraction efficiency and sensitivity by the membrane, such as temperature, membrane surface area, membrane thickness, geometry, sample volume, and sample flow rate, should be optimized for specific applications. Higher temperature has two opposite effects on the extraction efficiency. On the other hand, it facilitates mass transfer by increasing diffusion coefficient and, on the other hand, decreases analyte partition coefficient in the membrane. So, the temperature of the membrane module should be controlled to avoid fluctuation extraction efficiency and sensitivity. Another effective parameter

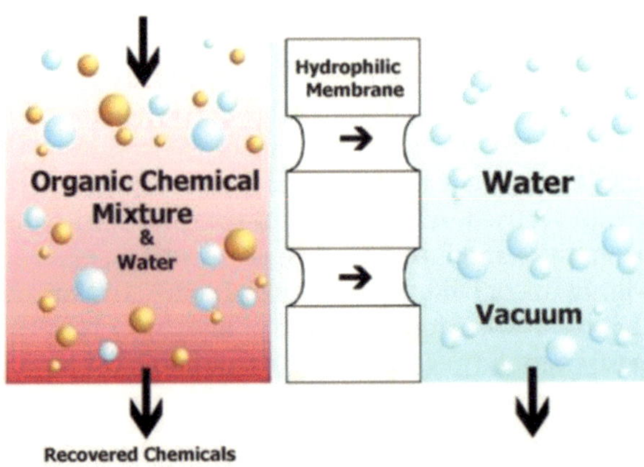

Figure 3.
The schematic diagram of membrane pervaporation.

is the membrane thickness. Faster mass transfer is achieved by using thinner membranes, and in the case of hollow fibers, using longer membranes and multiple fibers is better. Also by using the larger volume of the sample, higher sensitivity can be obtained. However, larger volumes take longer to extract, but this lower sample flow rate makes the extraction efficiency increase.

3. Concentration techniques for reducing the solvent volume

Once the analytes are diluted in the presence of a large volume of solvents during the extraction processes, they should be concentrated to analyze by instrumental methods as GC. If the amount of solvent to be removed is not very high and the analyte is nonvolatile, the solvent can be vaporized by a gentle stream of nitrogen gas flowing either across the surface or through the solution. But when a large volume of solvent should be removed, a rotary vacuum evaporator is used. In this case, the solution is placed in a round-bottomed flask which put in a heated water bath. A water-cooled condenser is attached at the top of flask to condense the evaporated solvent, and it distils into a separate container. Then, the flask is rotated continually to expose maximum liquid surface to evaporation. It should be noted that evaporation should stop before the solution reaches dryness.

For achieving smaller volume, e.g., less than 1 ml, a Kuderna-Danish concentrator is used. In this case, the solution is slowly heated in a warm water bath until the necessary volume is obtained. Also, an air-cooled condenser provides the solvent reflux [2].

4. Cleanup techniques

Sample cleanup is especially important for analytical separations such as GC, HPLC, and electrophoresis. Often, many solid matrices, as soil, biological materials, and natural products, contain hundreds of interferences at higher concentrations than those of the analytes. So, a cleanup step is vital to separate the trace amount of analyte from interferences. On the other hand, some high-boiling materials can cause a variety of problems such as the adsorption of analyte in the injection port or in front of a GC or HPLC column. Therefore, some positive and negative errors can be observed in the retention time of the analyte.

Some other cleanup techniques include gel permeation chromatography (GPC), acid-base partition cleanup, solid-phase extraction (SPE), and column chromatography, which are discussed in the following step [3].

Gel permeation chromatography (GPC) is a size-exclusion method which contained organic solvents (or buffers) and porous gels to separate macromolecules larger than analytes of interest. GPC is used to eliminate lipids, proteins, polymers, copolymers, natural resins, cellular components, viruses, steroids, and dispersed high-molecular-weight compounds from the sample. This method is suitable for both polar and nonpolar analytes [2]. On the other hand, GPC is usually used to remove high-boiling materials which condense in the injection port of a GC or the front of the GC column [28].

Acid-base partition cleanup is a liquid-liquid extraction procedure to separate acids such as organic acids and phenols from the base or neutral analytes like amines, aromatic hydrocarbons, and halogenated organic compounds, by adjusting pH. Also, this cleanup method is applied for petroleum waste prior to analysis [2].

Solid-phase extraction cartridge is a traditional column chromatography which is applied to clean up the biological, clinical, and environmental samples. Some of the SPE application examples are as follows [29]:

1. The cleanup of pesticide residues and chlorinated hydrocarbons.

2. The separation of nitrogen compounds from hydrocarbons.

3. The separation of aromatic compounds from an aliphatic-aromatic mixture.

4. The cleanup of steroids, esters, ketones, glycerides, alkaloids, and carbohydrates.

5. The cleanup of cations, anions, metals, and inorganic compounds.

As discussed in previous sections, the sufficient amount of a sorbent, which is loaded with the sample extract, has packed the SPE cartridge. Then, the analyte of interest is eluted through the column by an efficient eluting solvent, and the other contaminants are remained on the cartridge. The packing compound may be an inorganic material like either Florisil or one of many stationary phases which are commercially available [30].

5. Chemical derivatization analysis

Gas chromatography of volatile or nonpolar compounds may be done without derivatizing the sample; indeed, derivatives of compounds such as hydrocarbons or halogenated hydrocarbons cannot easily be prepared. It is possible to analyze polar compounds such as carboxylic acids and amines, without prior derivatization, on polar GC phases such as those based on polyethylene glycol. However, derivatization is useful in many instances where it may [31]:

1. Increase the volatility and decrease the polarity of polar compounds.

2. Stabilize compounds which are unstable at the temperatures required for GC.

3. Improve the separation of groups of compounds on GC column.

4. Yield information with regard to the number and type of functionalities present in mixtures of unknown compounds.

5. Improve the behavior of compounds toward selective detectors such as electron capture or nitrogen-selective detectors and mass spectrometry.

However, there are some drawbacks in using derivatization process before GC analysis:

1. The derivatizing agent may be difficult to remove and interfere in the analysis, and this is particularly disadvantageous when the purity of a compound is being assessed by GC.

2. The derivatization conditions may cause unintended chemical changes in a compound, for example, dehydration.

3. The derivatization step increases the time required for analysis.

For these reasons, GC with derivatization is less frequently employed in quality control applications, where the purity of a single substance or the components in a formulation are being determined.

Derivatization reactions are usually simple chemical reactions which are likely to occur in nearly quantitative yield such as acylation, alkylation, and silylation. In silylation reactions, some derivatives like trimethylsilyl (TMS) and tertiarybutyldimethylsilyl (TBDMS) can be prepared from a wide range of functional groups including hydroxyl, carboxylic, amine, amide, thiol, phosphate, hydroxide, and sulfonic. In acylation processes, acetate formation of the analytes is prepared by some derivatizing agents such as acetic anhydride, trifluoroacetic anhydride (TFAA), pentafluoropropionic anhydride (PFPA), and heptafluorobutyl anhydride (HFBA). Alkylation reactions may be used to derivatize carboxylic acids, amines, sulfonic acids, phosphonic acids, phosphates, barbiturates, uracils, purines, penicillins, thiols, and inorganic anions [31].

Derivatization reactions require relatively simple apparatuses:

1. Sample container and reaction vessel.

2. Heating and evaporation apparatuses.

3. Sample and reagent handling systems.

4. Removal systems for the exiting of the derivatizing reagents.

The certain standard procedures in derivatization are the following:

1. When volumes of reagent are small, reactions are carried out in 0.3 or 1 ml capacity Reacti-Vials or V-Vials. When volumes of solvents or reagents are greater, such as in aqueous phase reactions, then 3.5-ml screw-top sample tubes with aluminum-lined caps are used.

2. The reagents or solvent are evaporated under a stream of nitrogen gas with the sample maintained at 60–80°C in a heating block. Obviously, less volatile reagents require heating at higher temperatures for their efficient removal. If the sample is volatile, evaporation at a low temperature for a longer time may be required, or it may be better to inject it without removing the reagents.

3. Drying is carried out by passing the sample through ca. 3 cm of anhydrous sodium sulfate contained in a Pasteur pipette plugged with cotton wool. Anhydrous magnesium sulfate may be used too.

4. Dissolution of the derivatized sample prior to analysis is done by treating the sample with 2 ml of solvent for capillary column GC using the splitless injection mode (the volume may be adjusted if a split injection is used) or 100 μl for packed column GC. Since in most circumstances the derivatized compound should be clearly observed in relation to any interfering peaks from reagent residues, in injection 1 μl of product solution, 200 μg of material can be chosen as a proper starting point for the development of a method.

5. Removal of excess reagents is carried out by passage through a short column of Sephadex LH20. The sample is passed through a short column prepared by introducing Sephadex LH20 suspended in EtOAc/hexane (1:1, V/V) into a Pasteur pipette plugged with cotton wool and allowing the solvent to drain out to leave a pluge of ca. 3 cm of the adsorbent [31].

However, in some cases, the derivatization leads to sharper peaks and therefore to better separation and higher sensitivity. But the derivatization procedure requires more time and effort. Assadi et al. studied the determination of chlorophenols in water samples using simultaneous dispersive liquid-liquid microextraction and derivatization followed by gas chromatography-electron-capture detection [32]. In this research, dispersive liquid-liquid microextraction (DLLME) and derivatization coupled to gas chromatography-electron-capture detector (GC-ECD) was simultaneously applied for quantitative investigation of chlorophenols (CPs) in water sample. In this method, 500 μl of acetone, as disperser solvent, containing 10.0 μl of chlorobenzene, as extracting solvent, and 50 μl of anhydride acetic acid, as derivatizing reagent, was quickly injected into 5.00 ml of water sample containing CPs (analytes) and K_2CO_3 (0.5%, w/v) by a syringe. So, during a few seconds of time, the analytes were both derivatized and extracted simultaneously. Then, the mixture was centrifuged, and 0.50 μl of precipitated phase containing concentrated analytes was analyzed by GC-ECD instrument [32].

6. Superheated water extraction

When the temperature of liquid water is increased under pressure, between 100 and 374°C, its polarity is reduced significantly, and so, it can be applied as an extracting solvent for a wide variety of analytes. Its most interested application has been to determine PAHs, PCBs, and pesticides from environmental samples. Although it gives comparable results to Soxhlet extraction, the organic solvent consumptions have been significantly decreased, and quicker extractions were achieved. Unlike supercritical fluid extraction (SFE), unless the pressure is decreased and steam is applied, n-alkanes cannot be extracted. Other superheated water applications include the separation of required oils from plant substances where it preferably extracts the more important natural oxygenated compounds than steam distillation. The aqueous extract can be enriched via different methods such as solvent extraction, SPE, SPME, and extraction disk. On the other hand, the extraction can be coupled to LC or GC instruments, as online methods. In many cases the superheated water extraction is cleaner, faster, and cheaper than the conventional extraction methods [33].

The pressures, which are needed to keep a condensed state of water, are moderate in 15 bar at 200°C and 85 bar at 300°C. At any pressure, if the pressure falls below the boiling point of liquid water, superheated steam is produced. This superheated state possesses a significantly lower dielectric constant than that of the liquid state and also has gas-like diffusion velocity and viscosity properties. Consequently, superheated water behaves completely different from an extraction liquid solvent.

Superheated water has been widely used as an analytical extraction solvent. The changes in the polarity of water with increasing temperature have been also exploited in superheated water chromatographic methods [34].

Ozel et al. studied the analysis of volatile components from *Ziziphora taurica* subsp. *taurica* by steam distillation, superheated water extraction, and direct thermal desorption with GC·GC-TOFMS [35]. In this research, volatile compounds from the leaves of *Ziziphora taurica* subsp. *taurica* have been separated by steam distillation, superheated water extraction, and direct thermal desorption methods. The volatile constituents were analyzed by a perfect two-dimensional gas chromatography-time-of-flight mass spectrometry instrument. Some other researchers reported that superheated water is a powerful alternative extractor for separation of essential oils, because of its ability in working at low temperatures and obtaining higher speed extractions. Therefore, this makes the decomposition of volatile and heat-sensitive analytes be avoided. Extra advantages of the use of SWE are its simplicity, low cost, and friendly environment [36].

7. Single-drop microextraction

Single-drop microextraction (SDME) has witnessed incessant growth in the range of applications of sample preparation for trace organic and inorganic analysis. In SDME, a Teflon rod (or needle of a syringe) with a spherical recess at its one end is loaded with 8 μl of organic solvent (n-octane) containing the internal standard (n-dodecane) and immersed in aqueous sample taken in a 1 ml vial for a known period of time while being stirred. Thereafter, the rod is exited from the solution, and with a GC syringe, 1 μl of extract is injected into the GC column for analysis [37]. The stirrer rate of donor aqueous phase affects the solvent extraction speed and homogeneity of the obtained extract. SDME is comparable to SPME in terms of speed, precision, and sensitivity. But it is much cheaper than SPME and provides narrower peaks because in SDME, the solvent evaporation is faster than the analyte desorption from the fiber in SPME. However, in SDME, just little portion of extract is used to inject the GC column. By using a GC syringe instead of Teflon rod, the inconvenience of its filling can be eliminated. So, 1 μl of extract can be retracted back into the syringe after extraction process and injected directly into the GC column. Thus, the GC microsyringe can be used without any modification, and all other devices are general laboratory equipment. The GC microsyringe with a bend tip can hold the organic drop in place at controlled stirring rate. So, a number of instrumental analysis methods can be coupled to single-drop microextraction procedures.

There are two modules in SDME: direct immersion single-drop microextraction (DI-SDME) and headspace SDME. Their schematic diagrams are shown in **Figure 4**.

The direct immersion SDME is just applied for liquid samples containing non-polar or relatively polar analytes. To stabilize solvent drop during the extraction process, any insoluble and special materials must be removed from the sample medium, and a proper organic solvent with the least solubility in water, high boiling point, and high affinity to extract the analyte of interest should be chosen. Also at a moderated stirring rate, the drop must not be dislodged. However, DI-SDME is more favorable to match with GC method because of using water-immiscible solvent in the

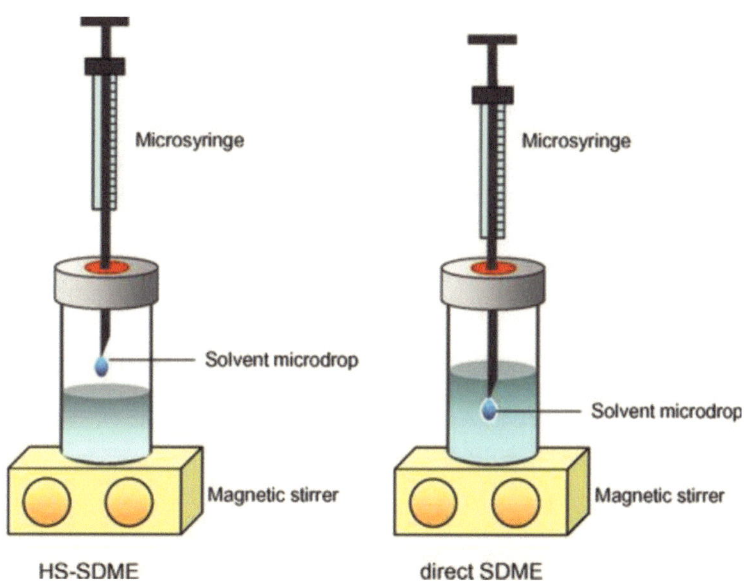

Figure 4.
Schematic diagrams of two modules in SDME.

drop. The searches are shown that N-octane and toluene possess the best extraction efficiency for nonpolar substances, while chloroform is more favorable to extract polar alkaloids, and then they can be analyzed by GC techniques [37]. One limitation of direct immersion SDME is the instability of the droplet at high stirring rates. Although high stirring rates enhance the extraction efficiency, to avoid the problem caused by elevated stirring speeds, a 1-μl microsyringe (instead of a more common 10-μl one) with some modification of its tip was used by Ahmadi et al. [38].

Ionic liquids have been established as alternative to organic solvents because of their high boiling point and viscosity which allow production of larger and more reproducible extraction drops. HPLC is a preferred method for analyzing ionic liquid extract, but their nonvolatility causes them unsuitable for GC analysis. To couple ionic liquid-based SDME to GC instrument, the extract is introduced via a removable interface which prevents entering of ionic liquid into the GC column, while the analytes can be entered quantitatively into the capillary column [37].

HS-SDME in which the organic droplet is held above the aqueous sample solution is most suitable for the consideration of volatile or semivolatile analytes [39]. The advantages of HS-SDME include the following: (1) Headspace SDME permits quick stirring of the sample solution with no concerning on the droplet stability. (2) The effects of nonvolatile matrix interferences are reduced, even if they are not eliminated. (3) In this mode, the analytes are distributed between three phases: the aqueous sample, headspace, and organic droplet. Since an elevated stirring rate of the sample solution enhances the mass transfer between the three phases. (4) In comparison with HS-SPME, HS-SDME shows to have the same precision and rate of analysis as HS-SPME. However, HS-SDME procedure possesses two special advantages over HS-SPME. At first, the approach of choosing solvents is wider. Second, the solvent cost (on the basis of several microliters) is negligible in comparison with the cost of commercially available fibers in SPME [39]. Alternatively, the use of SDME for headspace analysis seems relatively difficult, because of the requirement of the higher boiling point solvents. Although the most suitable solvents for gas chromatography should have relatively high vapor pressures or low boiling points, the limit of these solvents is obvious: they would evaporate too quickly in the headspace during extraction. Therefore, the select of suitable solvents should be the first decision in HS-SDME techniques.

8. Conclusions

Many methods are available for the treatment of volatile substances prior to instrumental analysis. In this chapter, the major methods which are leading to GC analysis have been explained. It has been observed that yet the classical techniques such as purge and trap, static headspace extraction, and liquid-liquid extraction act as important roles in chemical analysis of all sample types. New methods, such as SPME and membrane extraction, possess some advantages like convenient in automation and field sampling and reduction of solvent consumption, as well. If the analyst may be confronted with every difficulty, there is an appropriate available method to solve and face it. As a consequence, the main and primary analytical problem is to select the best sample preparation technique.

Acknowledgements

The author is appreciated of the support of Islamic Azad University, Shahr-e-Qods branch, Tehran, Iran.

Author details

Foujan Falaki
Department of Chemistry, Shahr-e-Qods Branch, Islamic Azad University, Tehran, Iran

*Address all correspondence to: foujanfalaki@gmail.com

References

[1] Skoog DA, Leary JJ. Principles of Instrumental Analysis. 4th ed. Saunders College Publishing; 1992

[2] Mitra S. Sample Preparation Techniques in Analytical Chemistry. Vol. 162. Hoboken, New Jersey: John Wiley & Sons, Inc.; 2003. pp. 1-439. ISBN: 0-471-32845-6

[3] Smith RM. Before the injection—Modern methods of sample preparation for separation techniques. Journal of Chromatography A. 2003;**1000**:3-27. DOI: 10.1016/S0021-9673(03)00511-9

[4] Shamsipur M, Falaki F, Shemirani F. Highly facile supported liquid membrane transport and removal of silver ion using dibenzyldiaza-18-crown-6 dissolved in a supramolecular solvent as selective ion carrier. Desalination and Water Treatment. 2016;**57**:25705-25717. DOI: 10.1080/19443994.2016.1162204

[5] Falaki F, Shemirani F, Shamsipur M. Surfactant-assisted transport of lead ion through a bulk liquid membrane containing dicyclohexyl-18-crown-6: Efficient removal of lead from blood serum and sea water. Journal of the Iranian Chemical Society. 2016;**13**(7):1257-1263. DOI: 10.1007/s13738-016-0839-5

[6] Cortada C, Vidal L, Tejada S, Romo A, Canals A. Determination of organochlorine pesticides in complex matrices by single-drop microextraction coupled to gas chromatography-mass spectrometry. Analytica Chimica Acta. 2009;**638**(1):29-35. DOI: 10.1016/j.aca.2009.01.062

[7] Handley AJ, editor. Extraction Methods in Organic Analysis. Sheffield: Sheffield Academic Press; 1999

[8] Lee H-B, Peart TE, Svoboda ML. Determination of endocrine-disrupting phenols, acidic pharmaceuticals, and personal-care products in sewage by solid phase extraction and gas chromatography-mass spectrometry. Journal of Chromatography A. 2005;**1094**(1-2):122-129. DOI: 10.1016/j.chroma.2005.07.070

[9] Stajnbaher D, Zupancic-Kralj L. Multiresidue method for determination of 90 pesticides in fresh fruits and vegetables using solid-phase extraction and gas chromatography- mass spectrometry. Journal of Chromatography A. 2003;**1015**:185-198. DOI: 10.1016/S0021-9673(03)01211-1

[10] Pawliszyn J. Solid-phase Micro-extraction: Theory and Practice. New York: Wiley-VCH; 1997

[11] Snow NH. Solid-phase micro-extraction of drugs from biological matrices. Journal of Chromatography A. 2000;**885**:445-455. DOI: 10.1016/S0021-9673(00)00192-8

[12] Fritz JS, Macka M. Solid-phase trapping of solutes for further chromatographic or electrophoretic analysis. Journal of Chromatography A. 2000;**902**:137-166. DOI: 10.1016/S0021-9673(00)00792-5

[13] Penton Z. Method development with solid phase microextraction. In: Scheppers Wercinski SA, editor. Solid Phase Microextraction: A Practical Guide. New York: Marcel Deker; 1999

[14] Goncalves C, Alpendurada MF. Solid-phase micro-extraction-gas chromatography-(tandem) mass spectrometry as a tool for pesticide residue analysis in water samples at high sensitivity and selectivity with confirmation capabilities. Journal of Chromatography A. 2004;**1026**:239-250. DOI: 10.1016/j.chroma.2003.10.117

[15] Yonamine M, Tawil N, Moreau RL d M. Ovandir Alves Silva, solid-phase micro-extraction-gas chromatography-mass spectrometry and headspace-gas chromatography of tetrahydrocannabinol, amphetamine, methamphetamine, cocaine and ethanol in saliva samples. Journal of Chromatography B. 2003;**789**:73-78. DOI: 10.1016/S1570-0232(03)00165-X

[16] Djozan D, Baheri T. Preparation and evaluation of solid-phase micro-extraction fibers based on monolithic molecularly imprinted polymers for selective extraction of diacetylmorphine and analogous compounds. Journal of Chromatography A. 2007;**1166**:16-23. DOI: 10.1016/j.chroma.2007.08.003

[17] Abdel-Rehim M. New trend in sample preparation: Online microextraction in packed syringe for liquid and gas chromatography applications I. Determination of local anaesthetics in human plasma samples using gas chromatography-mass spectrometry. Journal of Chromatography B. 2004;**801**:317-321. DOI: 10.1016/j.jchromb.2003.11.042

[18] Nakamura S, Daishima S. Simultaneous determination of alkylphenols and bisphenol A in river water by stir bar sorptive extraction with in situ acetylation and thermal desorption-gas chromatography-mass spectrometry. Journal of Chromatography A. 2004;**1038**:291-294. DOI: 10.1016/j.chroma.2004.03.040

[19] Liu W, Hu Y, Zhao J, Xu Y, Guan Y. Determination of organophosphorus pesticides in cucumber and potato by stir bar sorptive extraction. Journal of Chromatography A. 2005;**1095**:1-7. DOI: 10.1016/j.chroma.2005.07.107

[20] Quan L, Li S, Tian S, Xu H, Li A, Gu L. Determination of organochlorine pesticides residue in Ginseng root by orthogonal array design Soxhlet extraction and gas chromatography. Chromatographia. 2004;**59**:89-93. DOI: 10.1365/s10337-003-0120-9

[21] Goncalves C, Alpendurada MF. Assessment of pesticide contamination in soil samples from an intensive horticulture area, using ultrasonic extraction and gas chromatography-mass spectrometry. Talanta. 2005;**65**:1179-1189. DOI: 10.1016/j.talanta.2004.08.057

[22] Rissato SR, Galhiane MS, Knoll FRN, Apon BM. Supercritical fluid extraction for pesticides multiresidue analysis in honey: Determination by gas chromatography with electron-capture and mass spectrometry detection. Journal of Chromatography A. 2004;**1048**:153-159. DOI: 10.1016/j.chroma.2004.07.053

[23] Yusa V, Quintas G, Pardo O, Pastor A, de la Guardia M. Determination of PAHs in airborne particles by accelerated solvent extraction and large-volume injection-gas chromatography-mass spectrometry. Talanta. 2006;**69**:807-815. DOI: 10.1016/j.talanta.2005.11.018

[24] Liu R, zhou JL, Wilding A. Microwave-assisted extraction followed by gas chromatography-mass spectrometry for the determination of endocrine disrupting chemicals in river sediments. Journal of Chromatography A. 2004;**1038**:19-26. DOI: 10.1016/j.chroma.2004.03.030

[25] Grob K. Split and Splitless Injection in Capillary GC. 3rd ed. Heidelberg: Huthig; 1993

[26] Salemi A, Lacorte S, Bagheri H, Barcel´o D. Automated trace determination of earthy-musty odorous compounds in water samples by on-line purge-and-trap-gas chromatography-mass spectrometry. Journal of Chromatography A. 2006;**1136**:170-175. DOI: 10.1016/j.chroma.2006.09.087

[27] Kou D, San Juan A, Mitra S. Gas injection membrane extraction for fast on-line analysis using GC detection. Analytical Chemistry. 2001;**73**(22):5462-5467. DOI: 10.1021/ac010582i

[28] Pang GF, Cao YZ, Zhang JJ, Fan CL, Liu YM, Li XM, Jia GQ, Li ZY, Shi YQ, Wu YP, Guo TT. Validation study on 660 pesticide residues in animal tissues by gel permeation chromatography cleanup/gas chromatography–mass spectrometry and liquid chromatography–tandem mass spectrometry. Journal of Chromatography. 2006;**1125**:1-30

[29] Lopez-Avila V, Milanes J, Dodhiwala N, Beckert W. Cleanup of environmental sample extracts using florisil solid-phase extraction cartridges. Journal of Chromatography Science. 1989;**27**(5):109-215. DOI: 10.1093/chromsci/27.5.209

[30] Mills PA. Variation of florisil activity: Simple method for measuring absorbent capacity and its use in standardizing florisil columns. Journal of Association Off Analytical Chemistry. 1968;**51**:29

[31] Baugh PJ. Gas Chromatography: A Practical Approach. New York, United States: Oxford University Press Inc.; 1993

[32] Fattahi N, Assadi Y, Hosseini MRM, Jahromi EZ. Determination of chlorophenols in water samples using simultaneous dispersive liquid-liquid microextraction and derivatization followed by gas chromatography-electron-capture detection. Journal of Chromatography A. 2007;**1157**:23-29. DOI: 10.1016/j.chroma.2007.04.062

[33] Smith RM. Extractions with superheated water. Journal of Chromatography A. 2002;**975**:31-46. DOI: 10.1016/S0021-9673(02)01225-6

[34] Pawlowski TM, Poole CF. Solvation characteristics of pressurized hot water and its use in chromatography. Analytical Communications. 1999;**36**:71-75. DOI: 10.1039/A809684H

[35] Ozel MZ, Hamilton FG, JF, Lewis AC. Analysis of volatile components from Ziziphora taurica subsp. Taurica by steam distillation, superheated-water extraction, and direct thermal desorption with GC•GC-TOFMS. Analytical Bioanalytical Chemistry. 2005;**382**(1):115-119. DOI: 10.1007/s00216-005-3156-x

[36] Ozel MZ, Clifford AA. Superheated water extraction of fragrance compounds from Rosa canina. Flavour and Fragrance Journal. 2004;**19**(4):354-359. DOI: 10.1002/ffj.1317

[37] Jain A, Verma KK. Recent advances in applications of single-drop microextraction: A review. Analytica Chimica Acta. 2011;**706**(1):37-65. DOI: 10.1016/j.aca.2011.08.022

[38] Ahmadi F, Assadi Y, Hosseini SMRM, Rezaee M. Determination of organophosphorus pesticides in water samples by single drop microextraction and gas chromatography-flame photometric detector. Journal of Chromatography A. 2006;**1101**(1-2):307-312. DOI: 10.1016/j.chroma.2005.11.017

[39] Xu L, Basheer C, Lee HK. Developments in single-drop microextraction. Journal of Chromatography A. 2007;**1152**(1-2):184-192. DOI: 10.1016/j.chroma.2006.10.073

Section 4

Application

Chapter 4

Trends for Pyrazole Fragmentation Determined by Gas Chromatography Coupled with Mass Spectrometry

Clarissa Piccinin Frizzo, Bruno Luís Hennemann, Bruna Luisa Kuhn, Keli Maiara Wust, Alisson Vasques Paz, Marcos Antonio Pinto Martins, Nilo Zanatta and Hélio Gauze Bonacorso

Abstract

In this chapter, we present a review on pyrazole fragmentation by gas chromatography coupled with mass spectrometry, in order to evaluate the substituent effect on pyrazole fragmentation. Our objective was to present a comprehensive study on the fragmentation pattern of substituted pyrazoles, contribute to the systematization of knowledge, and offer support to researchers in the characterization of pyrazoles via a comprehensive and versatile technique such as gas chromatography coupled with mass spectrometry. The pyrazole fragmentation showed two important processes: (i) expulsion of HCN from $[M]^{+\bullet}$ and $[M–H]^{+}$ and (ii) the loss of N_2 from $[M–H]^{+}$. Substituents such as D, Me, Br, Cl, and Ph did not influence these two processes; however, the presence of nitro, acetyl, oxime, diphenyl, or methyl and nitro in the *ortho*-position transforms the two processes into secondary fragmentation or results in their absence in the fragmentation of the said pyrazoles.

Keywords: pyrazoles, mass fragmentation, low-resolution spectrum, methyl, nitro

1. Introduction

The pyrazole is one of the most important heterocycles. It is a versatile five-membered heterocycle and a lead molecule in pharmaceutical development, due to its wide range of biological activity (e.g., antimicrobial, anticancer, cytotoxic, analgesic, anti-inflammatory, antihypertensive, antiepileptic, and antidepressant activities) [1]. Some drugs containing pyrazoles (e.g., celecoxib, novalgin, ramifenazone, fipronil, rimonabant, and pyrazofurin) are already in the market. Because of this, research on the synthesis, structural characterization, and properties of pyrazoles is ever increasing. Currently, a search—using pyrazole as topic—on the Web of Science showed 14,000 results, 327 of which are reviews. Early works on the mass spectrum fragmentation pattern of pyrazoles date from the early 1970s. Over approximately one decade, a few scientists—including Thuijl et al. [2–6],

Bowie et al. [7, 8], and Finar and Millard [9]—investigated the mass fragmentation of substituted pyrazoles. In 2005, Santos et al. published a study focused on the mass fragmentation of substituted pyrazoles. Advances in the use, as an analytical tool, of gas chromatography coupled with mass spectrometry—including ionization modes such as electron spray and chemical ionization—and the use of high-resolution mass spectrometry led to novel structural elucidation, fragmentation mechanisms, and other properties for pyrazoles. However, up until 2005, pyrazole fragmentation, as well as the substituent effect, had not been studied from the perspective of these advances. The aforementioned analytical tool brings new insights regarding structural and other properties of pyrazoles. Because of this, and considering that our objective in this chapter is to present a comprehensive study on the fragmentation pattern of substituted pyrazoles, contribute to the systematization of knowledge, and assist researchers in the characterization of pyrazoles via a comprehensive and versatile technique such as gas chromatography coupled with mass spectrometry, only the works published up until 2005 will be discussed here. Thus, in this chapter, we will present the studies on pyrazole fragmentation by gas chromatography coupled with mass spectrometry, in order to evaluate the substituent effect on pyrazole fragmentation and to present a comprehensive study on the fragmentation pattern of substituted pyrazoles.

2. Trends in mass spectrum fragmentation of pyrazoles

Thuijl et al. [2] studied the fragmentation of a series of 10 pyrazoles—substituted at the 2,3,4-, and 5-position—at low resolution. It is worth noting that most pyrazoles are unsubstituted at the 1-position (*NH*-pyrazole); therefore, they exist as a tautomeric equilibrium of two forms (**Figure 1**).

The molecular structure of the *NH*-pyrazoles investigated by Thuijl et al. [2] is shown in **Table 1**. In general, the substituent at the 4-position was changed in the pyrazole series, and the compounds are divided into the following two groups in accordance with the substituent: (i) chloro- and bromine-substituted pyrazoles, including mono-, di-, and tri-substituted ones, and (ii) cyano-, nitro-, carboxy-, and

Figure 1.
Representation of the two tautomers of the NH-pyrazoles.

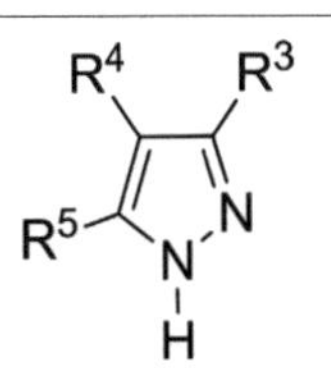

Comp.	R^1	R^3	R^4	R^5	Comp.	R^1	R^3	R^4	R^5
1	H	H	H	H	**6**	H	H	CN	H
2	H	H	Cl	H	7	H	H	NO_2	H
3	H	H	Br	H	**8**	H	H	$COCH_3$	H
4	H	H	Br	Br	**9**	H	H	C_6H_5	H
5	H	Br	Br	Br	10	CH_3	H	H	H

Table 1.
*Molecular structure of NH-pyrazoles **1–10**.*

phenyl-substituted pyrazoles. The authors included an unsubstituted pyrazole and a 1-methyl substituted pyrazole in the series, in order to evaluate the effect that this position has on the fragmentation trend of 4-substituted *NH*-pyrazoles. The simplest pyrazole analyzed was the unsubstituted pyrazole (**1**). The authors showed that the fragmentation follows two distinct routes. The predominant feature is expulsion of HCN (*b*, *m/z* 41, $[C_2H_3N]$) or H (*a*, *m/z* 67) from the $[M]^{+\bullet}$ ion or expulsion of HCN from [M–H] (*c*, *m/z* 40, $[C_2H_2N]$)—see **Figure 2**.

The second process involves the loss of a nitrogen from [M–H] (*a*, *m/z* 67), which furnishes the cyclopropenyl ion (*e*, *m/z* 39, $[C_3H_3]^+$). The intensity ratio of the ions formed by the loss of HCN ($b + c + d + f$) and loss of N_2 ($e + g + [C_3H]^+ + [C_3]^{+\bullet}$) is 5:1. Elimination of acetylene also occurs but only to a very small extent. Apart from the loss of HCN and $H^\bullet$ from the molecular ion, Khmel'nitskii et al. [10] also reported on the loss of N_2 and HCN from [M–H], which leads to $[C_3H_4]^{+\bullet}$ (*e*, *m/z* 39) and $[C_2H_2N]^+$ (*c*, *m/z* 40), respectively—this was not identified in the study of Thuijl et al. [2]. High-resolution measurements did not reveal the presence of $[C_3H_4]^{+\bullet}$. Metastable defocusing indicated that $[C_2H_2N]^+$ (*c*, *m/z* 40) has only two precursors: *b* and *a*. The transitions 39 → 38 can be assigned to two processes: $e \rightarrow g$ and $d \rightarrow f$—see **Figure 2**. The latter process was observed in the spectrum of 3(5),4-dibromopyrazole **4** (see **Figure 4**), in which *e*, m/z 39 was absent.

4-Chloro- (**2**) and 4-bromopyrazole (**3**)—see **Figure 3**—have similar fragmentation. Although $[M\text{–}H]^{+\bullet}$ is virtually absent, the loss of $Br^\bullet$ (*a*, *m/z* 67) is followed by expulsion of N_2 to give *e*, *m/z* 39. Two successive losses of HCN from $[M]^{+\bullet}$ lead to $[CH_3Br]^{+\bullet}$ (*h*, *m/z* 92, 94) in preference to $[CH_2]^{+\bullet}$ (*h*, *m/z* 14), which is also weak in unsubstituted pyrazoles. The expulsion of HCN and radical bromine was also observed in 3,4-dibromo (**4**). However, for pyrazole **4**, the $a \rightarrow e$ process (loss of N_2) was absent. Instead of $[C_3H_2N_2Br]^+$ being formed, *a* lost the second $Br^\bullet$ to give $[C_3H_2N_2]^{+\bullet}$ (*i*, *m/z* 66), which formed $[C_2HN]^{+\bullet}$ (*d*, *m/z* 39) after elimination of HCN (**Figure 4**).

The mass spectrum of pyrazole **5** showed that the loss of HCN is still important for tribrominated pyrazoles. The formation of $[C_2NBr_3]^{+\bullet}$ (*b*, *m/z* 275–281) involves a bromine migration similar to that reported by Bowie et al. [7] for tribromoimidazole. Another example of a bromine migration is the formation of $[CBr_2]^{+\bullet}$

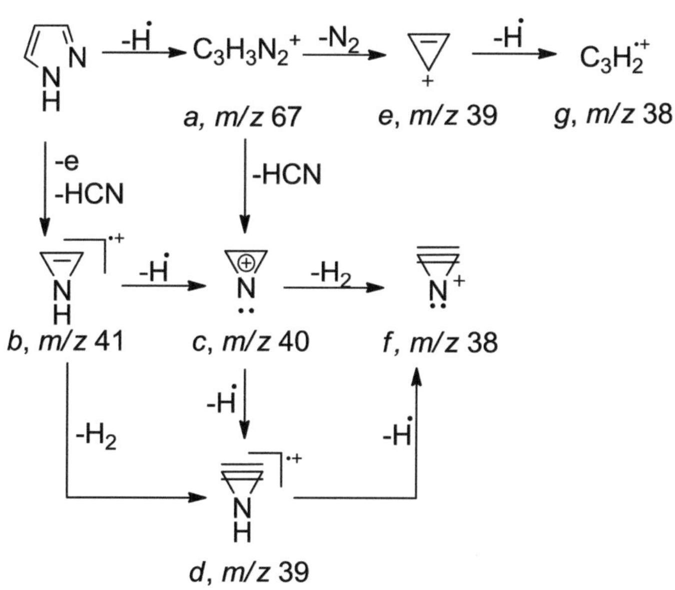

Figure 2.
Principal fragmentation of unsubstituted pyrazole 1.

(*h*, *m/z* 170–174) from $[C_2HNBr_2]^{+\bullet}$ (*b*, *m/z* 197–201) in **4**. Principal mass fragmentation of compounds **4** and **5** is shown in **Figure 5**.

In 4-cyanopyrazole **6**, the elimination of HCN plays such an important role that $[M–CN]^{+\bullet}$ (*f*, *m/z* 66) is formed, which breaks down in the same way as in **3** and **4**. Principal mass fragmentation of **6** is shown in **Figure 6**.

The mass spectrum of 4-nitropyrazole 7 shows the usual fragments associated with nitro groups in aromatic compounds: $[M–O]^{+\bullet}$ (*a*, *m/z* 97), $[M–NO]^+$, and $[M–NO_2]^+$. Loss of the $–NO_2$ substituent gives *a*, *m/z* 67, which forms subsequent fragments—see **Figure 2**. However, another route to *c*, $[C_2H_2N]^+$ was found for 7—see **Figure 7**.

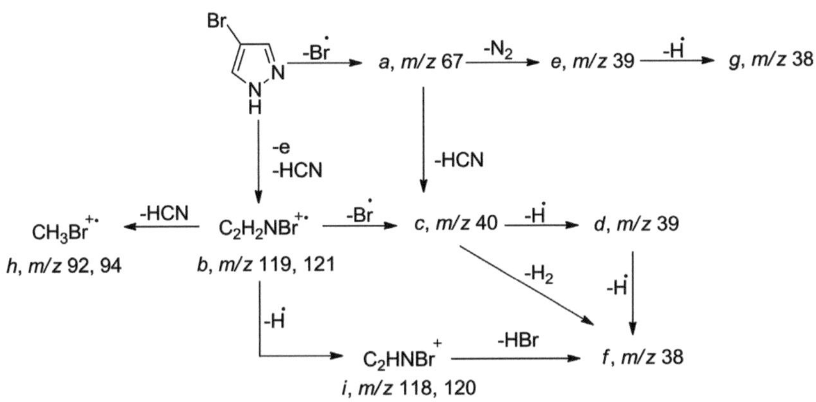

Figure 3.
Principal mass fragmentation of 4-bromopyrazole 3.

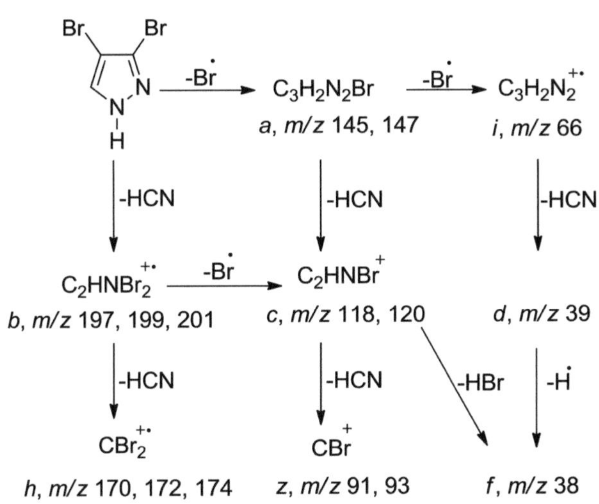

Figure 4.
Principal mass fragmentation of 3,4-dibromopyrazole 4.

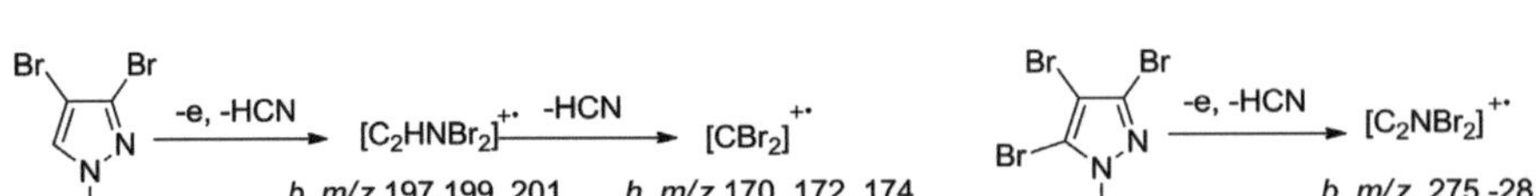

Figure 5.
Principal mass fragmentation of compounds 4 and 5.

The behavior of 4-acetylpyrazole **8** indicates that $[M–CH_3]^+$ (*j*, *m/z* 95) is the base peak and that the loss of the substituent is a two-step process (**Figure 8**). It should be noted that the relative intensity of the unfavorable species $[C_3H_3N_2]^+$ (*a*, *m/z* 67) is small compared to that of $[C_3H_2NO]^+$ (*k*, *m/z* 68), formed by the HCN lost, which is stabler. In contrast to the previous compound, no elimination of HCN from the molecular ions occurs.

The mass spectrum of 4-phenylpyrazole **9** closely resembles that of the unsubstituted pyrazole. Loss of HCN from $[M]^{+\bullet}$ leads to $[C_8H_7N]^{+\bullet}$ (*b*, *m/z* 117) and $[C_8H_6N]^+$ (*c*, *m/z* 116). Alternatively, successive losses of $H^\bullet$ and N_2 lead to the phenylcyclopropenyl ion $[C_9H_7]^+$ (*e*, *m/z* 115). Expulsion of HCN from *b* and *c* gives (*f*, *m/z* 90) $[C_7H_6]^{+\bullet}$ and (*d*, *m/z* 89) $[C_7H_5]^+$, respectively (**Figure 9**).

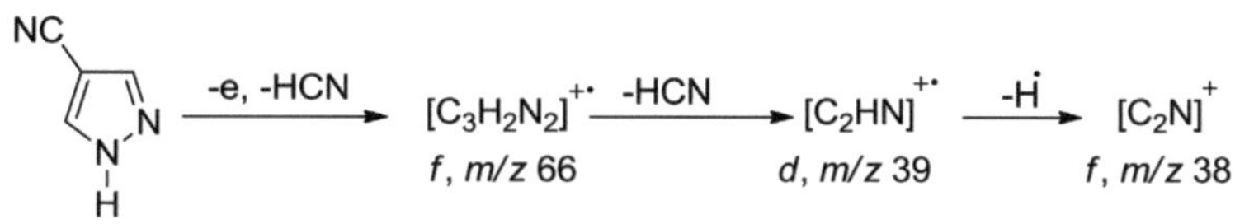

Figure 6.
Principal mass fragmentation of compound **6**.

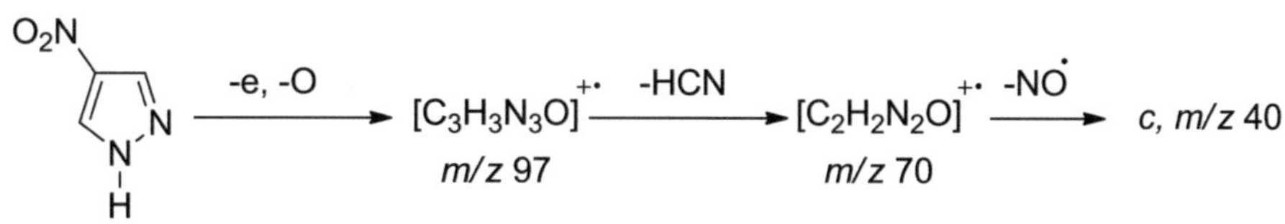

Figure 7.
Principal mass fragmentation of compound **7**.

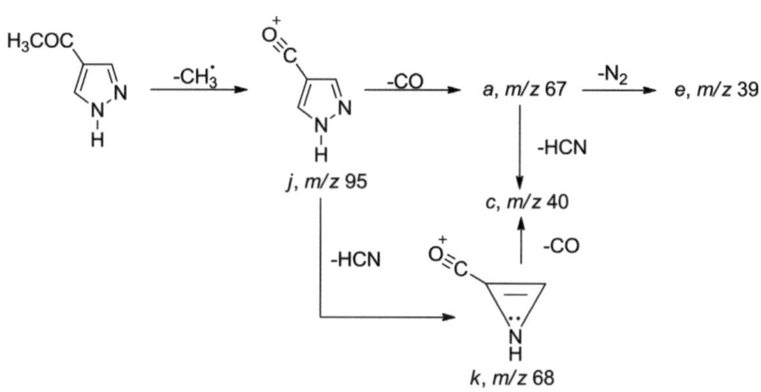

Figure 8.
Principal mass fragmentation of compound **8**.

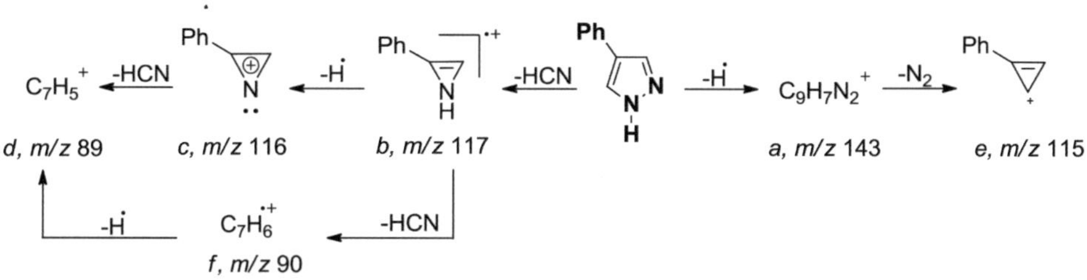

Figure 9.
Principal mass fragmentation of compound **9**.

The low-resolution spectrum of 1-methylpyrazole **10** showed that a H• radical is lost preferably from the methyl group, which may be rearranged to give *m* (*m/z* 81)—see **Figure 10**. The fact that the ion $[C_4H_5]^+$ (*e*, *m/z* 53) is formed by elimination of N_2 from $[M–H]^+$ means that it does not exclusively rearrange to the pyrimidinium ion but also, to some extent, to the pyridazinium ion (*m*, *m/z* 81) (**Figure 10**). In the case of H• abstraction from C-5, another possible explanation for the loss of N_2 from $[M–H]^+$ is migration of the methyl group to C-5 (*n*, *m/z* 81) prior to formation of the methylcyclopropenyl ion (*e*, *m/z* 53). Although (*n*, *m/z* 81) might be expected to expel acetonitrile, an *m/z* 81 → *m/z* 40 transition was not observed.

In conclusion, Thuijl et al. [2] showed that pyrazole and its 4-substituted derivate have stable molecular ions, which generally form the base peak. A notable exception is 4-acetylpyrazole, for which the $[M–CH_3]^+$ ion is the base peak. Two important processes are observed in unsubstituted pyrazoles: the expulsion of HCN from $[M]^{+\bullet}$ and $[M–H]^+$ and the loss of N_2 from $[M–H]^+$—the former process is the most predominant. Substitution influences the relative importance of these two processes. Whereas in 4-bromopyrazole the patterns remain essentially unchanged, introduction of more bromine atoms makes initial loss of HCN less important, due to the competing loss of bromine radicals. However, secondary loss of HCN remains important, resulting in the suppression of the expulsion of N_2. The influence of the 4-nitro and 4-acetyl substituents is much more pronounced, because the substituents can be in more than one step, giving rise to the stable ions $[M–O]^{+\bullet}$ and $[M–NO]^+$ in the former and $[M–CH_3]^+$ in the latter. In these compounds, the loss of HCN, though still strong as a secondary step, is no longer a primary process. On the other hand, the 4-cyano and 4-phenyl substituents promote the loss of HCN in such a manner that the ion formed by the loss of the substituent is practically absent. In 4-cyanopyrazole, the initial loss of HCN is very strong, resulting in an ion with *m/z* 66, while in 4-phenylpyrazole, two consecutive losses of HCN occur. This marked influence of the substituent makes it difficult, in some cases, to establish the presence of a pyrazole compound from a low-resolution spectrum.

Thuijl et al. [3] reported the fragmentation of the following: monodeuterated pyrazoles at the 1-position (**11**) and 4-position (**12**); dideuterated pyrazoles at the 1,4_position (**13**) and 3,5-position (**14**); and trideuterated pyrazoles at the 1,3,5-position (**15**) at low resolution—see **Table 2**. The fragmentation of these pyrazoles was similar to the simplest pyrazole (**1**), in that two of the most important fragmentations involved the loss of H• and HCN from the molecular ion. Similarly, these results can be explained in terms of cleavage of a C–H bond in an α-position relative to a nitrogen atom. According to the authors, this α-cleavage is a frequently

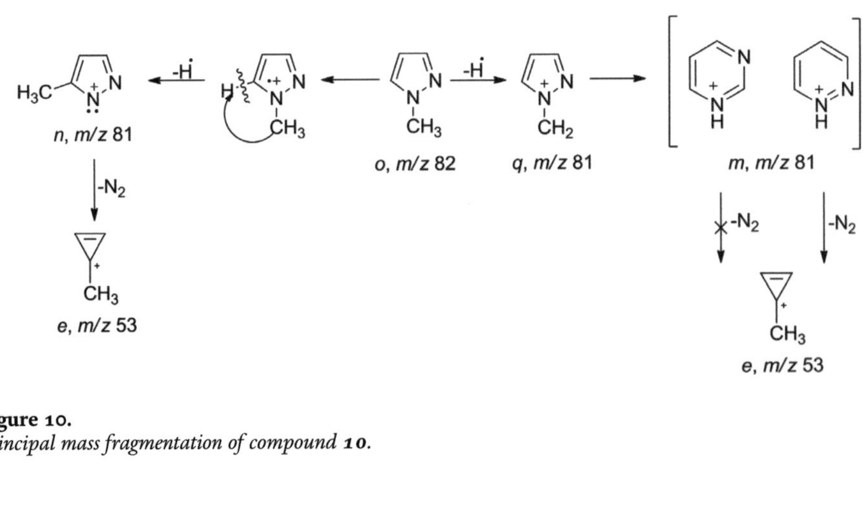

Figure 10.
Principal mass fragmentation of compound 10.

encountered phenomenon in mass spectra of compounds possessing heteroatoms. The loss of HCN from the molecular ion of non-deuterated pyrazoles gives rise to a peak corresponding to the formation of $C_2H_3N^+$ (*m/z* 41). The peak at *m/z* 40 ($C_2H_2N^+$) arises because of the subsequent loss of H^+. In a monodeuterated compound, the loss of HCN and DCN forms fragments *m/z* 42 and *m/z* 41, respectively. The mass spectra of the 3,5-dideuterated pyrazole (**14**) and 1,3,5-trideuterated pyrazole (**15**) showed that position 3(5) is favored, but the mass spectra of the other three deuterated compounds do not, a priori, exclude contributions from the other two positions. The authors concluded that the loss of HCN from the molecular ion of pyrazole occurs from the 3(5) position with a high specificity. The authors also noted that more quantitative and high-resolution spectra would be useful to enable a complete understanding of the fragmentation of deuterated pyrazoles.

Aldous and Bowie [8] also investigated deuterated 3,5-diphenylpyrazoles at low resolution (**Table 3**). The fragmentation of 3,5-diphenylpyrazoles produced $[C_{13}H_9]^+$ ions through the elimination of $N_2H^\bullet$ and $^\bullet C_2H_2$ from M^+. This fragmentation differs from other diphenylpyrazoles, imidazoles, and isoxazoles. For example, fragmentation of 4,5-diphenylpyrazole forms $[C_{13}H_9]^+$ ions (*m/z* 165) from molecular ions, through the [M–HCN–HCN–H$^\bullet$] or [M–$C_2H_3N_2^\bullet$] process. Another usual feature of the 4,5-diphenylpyrazole spectra is the reversible hydrogen rearrangement, which occurs between the N-*H* and the *ortho*-hydrogens of the phenyl rings.

The molecular ions of 3,5-diphenylpyrazoles decompose through the loss of N_2D and N_2H, with N_2H forming at higher intensity. This result parallels the reversible hydrogen rearrangement observed for 4,5-diphenylpyrazoles. The reversible transfer of a phenyl hydrogen to nitrogen followed by back transfer of either $H^\bullet$ or $D^\bullet$ occurs prior to or during the initial elimination ($N_2D^\bullet$ and $N_2H^\bullet$ from **16**), and as the lifetime of the decomposing molecular ion increases, H/D scrambling in the phenyl rings competes with the reversible transfer reaction. Proof of the D transfer to the phenyl substituents is substantiated by the observation of the two processes $[M]^{+\bullet} \rightarrow [C_6H_5]^+$ and $[M]^{+\bullet} \rightarrow [C_6H_4D]^+$. The spectra of **16** and **17** show that there is no exchange between the 1- and 4-H/D atoms or between the 4-H (or D) and the

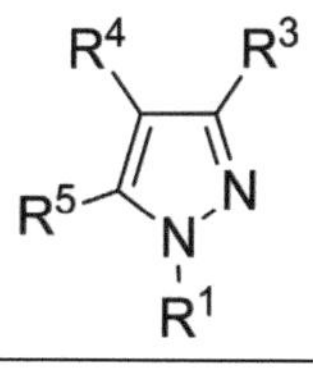

Comp.	R^1	R^3	R^4	R^5	Comp.	R^1	R^3	R^4	R^5
1	H	H	H	H	**13**	D	H	D	H
11	D	H	H	H	**14**	H	D	H	D
12	H	H	D	H	**15**	D	D	H	D

Table 2.
*Mass spectral investigation of compounds **1** and **11–15**.*

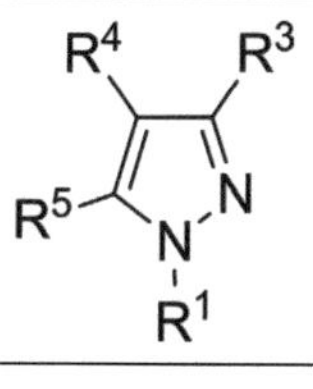

Comp.	R^1	R^3	R^4	R^5
16	D	Ph	H	Ph
17	D	Ph	D	Ph
18	H	Ph	H	Ph

Table 3.
*Structure of pyrazoles **16–18**.*

phenyl hydrogens. The metastable decompositions of **16**, **17**, and **18** show that the $[M^{+}– N_2H^{\bullet}]^{+}$ ion from 3,5-diphenylpyrazole eliminates $^{\bullet}C_2H_2$ after complete carbon and hydrogen scrambling. The other fragmentations shown in **Figure 11** are unexceptional and occur without prior equilibration of the N-*H* and phenyl hydrogens.

Finar and Millard [9] studied the low-resolution mass spectrometry of 1-phenylpyrazole-4-yl-oximes (**19–23**)—see **Table 4**. This is a typical substrate of the Beckmann rearrangement for corresponding 4-amido-1-phenylpyrazoles. If the ions of the anti-oximes were to undergo the Beckmann rearrangement for the corresponding 4-amido-1-phenyl-pyrazole, the $C_{10}H_7N_2O^{+}$ ion (*m/z* 171) formed by the breaking of the CO–NH bond would be present (**Figure 12**).

Aliphatic aldoximes, such as 4-formyl-1-phenyl-pyrazole oxime (**19**)—**Figure 12**, have relatively intense molecular ions. The spectrum is characterized by the elimination of water from the molecular ion (*a*, *m/z* 169), as the loss of a hydroxyl radical (*b*, m/z 170). Oxygen is also lost to a small extent (*c*, *m/z* 171). The base peak represents the oxime lost followed by the formation of the 1-phenylpyrazole ion (*d*, *m/z* 144). An appropriate metastable ion is formed directly from the molecular ion by a process which must involve the breaking of the 4-substitution and

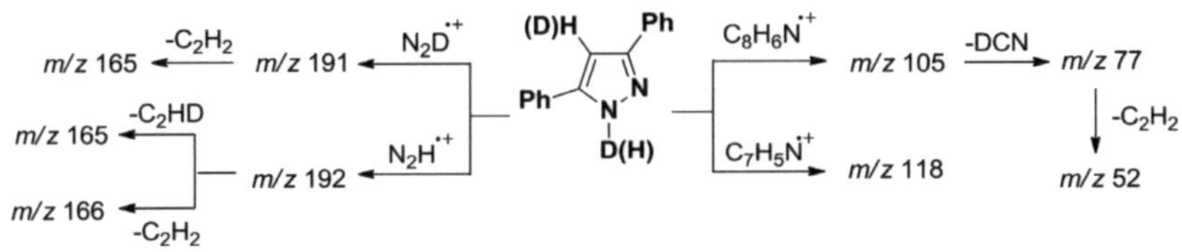

Figure 11.
Fragmentation pattern of compounds 16–18.

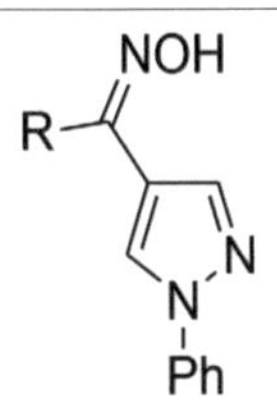

Comp.	R	Comp.	R
19	H	22	*i*-pr
20	Me	23	*t*-Bu
21	Et	24	Ph

Table 4.
Mass spectral investigation of compounds 20–24.

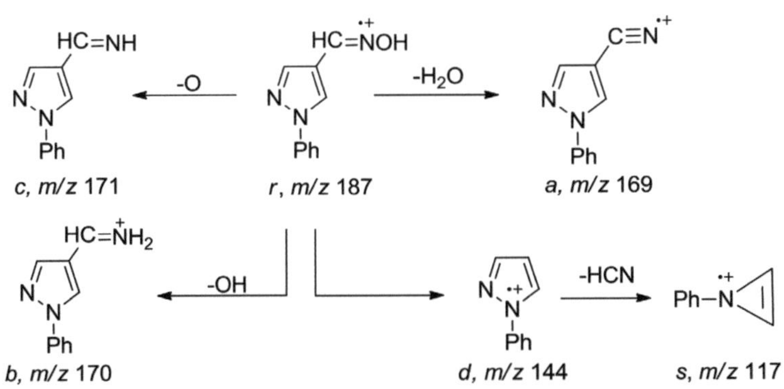

Figure 12.
Mass fragmentation of aldoxime pyrazole 19.

simultaneous transfer of hydrogen to the pyrazole ring. The substitution of this hydrogen by deuterium causes the ion to move to *m/z* 145, thus confirming that the hydrogen is transferred to the pyrazole.

For the ketoxime pyrazoles **20–23**, the one-step elimination of water in **19** is not paralleled by a corresponding one-step loss of the appropriate alcohol, although the elements of alcohol are lost as a hydroxyl radical followed by the alkyl group. Again, exchange of the oximino-hydrogen for deuterium causes this ion to move to *m/z* 145, which indicates that the mechanism described in **Figure 12** is present here. The loss of oxygen from the molecular ion remains a significant process but becomes less so as the substituent increases in size, possibly because of the increased number of fragmentation modes available. **Figure 13** shows how a typical ketoxime—ethyl-1-phenylpyrazole-4-yl oxime **21**—fragments. The ion of *m/z* 170 that arises from the loss of ethylene—perhaps via a McLafferty-type rearrangement—is also present in the spectra, again being formed from M–OH ions. When R = *i*-pr (**22**) and *t*-bu (**23**), it appears that the alkyl substituents undergo rearrangement to the linear-chain isomers in the molecular ions, because **22** eliminates an ethylene molecule and **23** eliminates a propene molecule. Such rearrangements have been noted previously in the mass spectra of branched alkyl compounds [11].

The 4-benzoyl-1-phenyl-pyrazole oxime **24** behaves differently to the other compounds—see **Figure 14**. The molecular ion at *m/z* 263 and the rearrangement ion at 144 are by far the most intense ions in the spectrum. Besides the anticipated loss of oxygen and a hydroxyl radical from the molecular ion, the unexpected ejection of NO, HNO, and H_2NO occurs. The M–H_2NO ion at *m/z* 231 fragments further—due to the loss of HCN—and yields an ion encountered in the mass spectrum of benzophenone oxime. It appears that the molecular ions of these oximes do not undergo the Beckmann rearrangement, but rather a rearrangement of the oximino-hydrogen atom, which has not yet been encountered in the mass spectra of oximes.

Luijten and Thuijl [4] studied how the presence of methyl and nitro groups in the pyrazole ring affects the fragmentation of these compounds in the low-resolution spectrum (**Table 5**).

For compounds **25–27**, the common characteristics are the initial loss of $NO^•$ (which leads to the formation of *o*, *m/z* 97) followed by the loss of $O^•$ or the loss of $NO_2^•$, which leads to the ion *m/z* 81 (**Figure 15**). The difference between the three compounds occurs due to the fragmentation of the ion *m/z* 81 (*p* to **25** and **27**, and *q* to **26**)—see **Figure 15**. Although all compounds have the *r*, *m/z* 53 and *s*, *m/z* 54 ions, their relative abundance is different. At **25** and **27**, the *r*, *m/z* 53 is the most

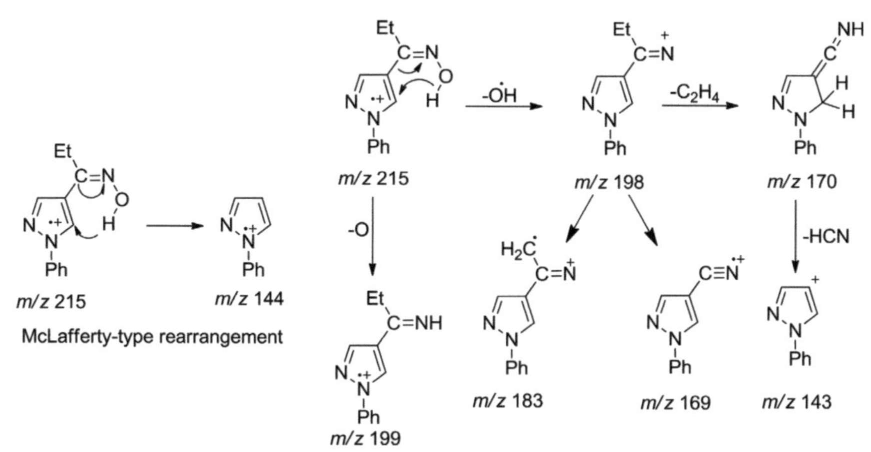

Figure 13.
Mass fragmentation of ketoxime pyrazoles 20–23.

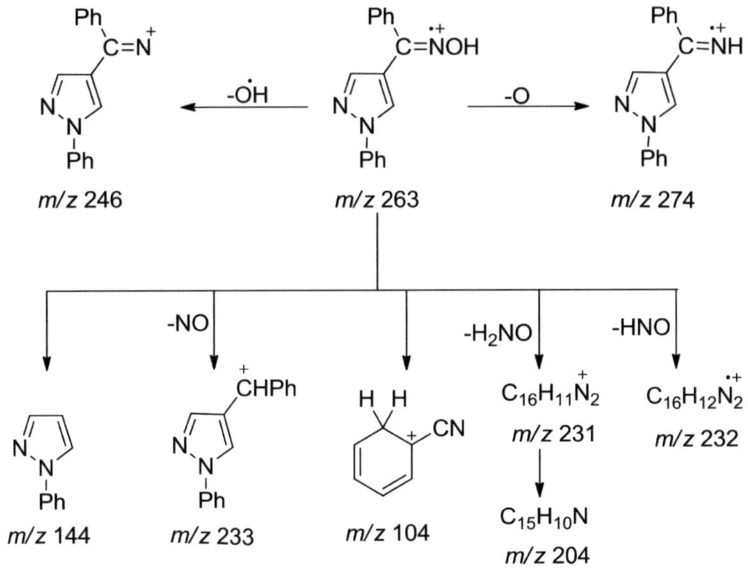

Figure 14.
4-Benzoyl-1-phenyl-pyrazole oxime 24.

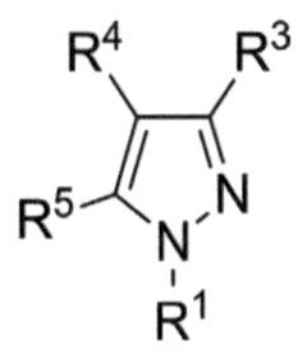

Comp.	R^1	R^3	R^4	R^5	Comp.	R^1	R^3	R^4	R^5
25	NO_2	CH_3	H	H	**30**	CH_3	H	H	NO_2
26	NO_2	H	CH_3	H	**31**	H	CH_3	H	NO_2
27	NO_2	H	H	CH_3	**32**	H	NO_2	CH_3	H
28	CH_3	NO_2	H	H	**33**	H	CH_3	NO_2	H
29	CH_3	H	NO_2	H					

Table 5.
Structure of pyrazoles 25–33.

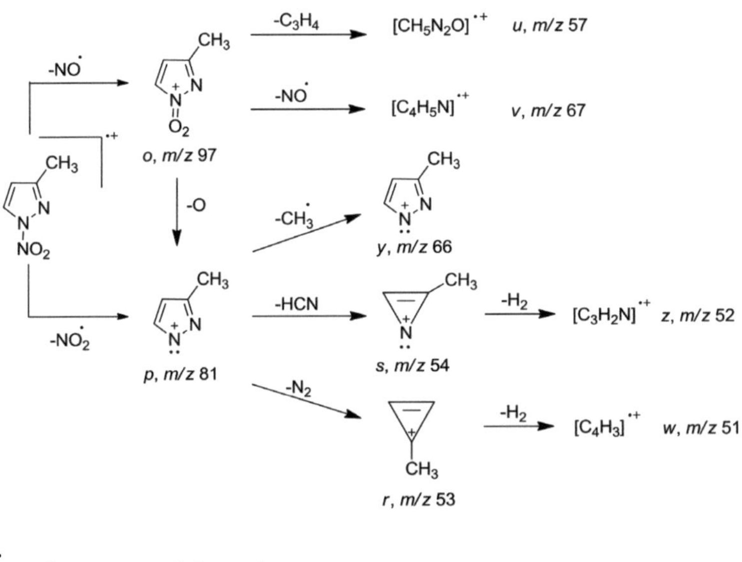

Figure 15.
Fragmentation of 1-nitro-3-methylpyrazole 25.

abundant ion, which is expected, because the two compounds have equivalent structures, as can be seen in **Figure 16**, while for compound **26**, the most abundant ion is *s*, *m/z* 54. **Figure 16** explains the production of *r*, *m/z* 53 and *s*, *m/z* 54 for nitropyrazoles. The formation of the *s*, *m/z* 54, occurs due to the loss of HCN, whereas the formation of *r*, *m/z* 53 is due to the loss of N_2. Luijten and Thuijl [4] attributed these differences to the position of the methyl groups in pyrazoles, because for **25** and **27**, *r*, *m/z* 53 is stabler, while for **26**, *s*, *m/z* 54 is stabler. Fragmentation of compound **25** is explained in **Figure 15**.

Of note in the mass spectrum of compound **28** is the very stable molecular ion. Different from pyrazoles **25–27**, the ion *g*, *m/z* 81 has low abundance, which suggests the formation of ion *h*, *m/z* 80 through the loss of HNO_2 or a loss of $H^{\bullet}$ from the $[M–NO_2]^+$ fragment. The presence of ions *b*, *m/z* 97 and *f*, *m/z* 111 indicates the presence of the nitro group in the molecular ion. Upon fragmentation, the ion *b*, *m/z* 97 generates the ion *c*, *m/z* 69, which can fragment in two distinct ways—one generating the ion *d*, *m/z* 43 through the release of -C_2H_2 and the other generating the ion *e*, *m/z* 42 through the release of -HCN. Another interesting fragmentation generates the ion *d*, *m/z* 43 directly from the molecular ion, through the loss of $-C_3H_2NO_2^{\bullet}$. The fragments are summarized in **Figure 17**.

The fragmentation of 1-methyl-4-nitropyrazole **29** (see **Figure 18**) was different from the 3-isomer **28**. Starting from ion *b*, *m/z* 97, Luijten and Thuijl [4] observed only the formation of ion *e*, *m/z* 42, whereas in isomer **28**, starting from the same fragment, they observed the formation of ions *m/z* 42 and *m/z* 43. The formation of the ion *e*, *m/z* 42 was also observed from ion fragmentation *d*, *m/z* 111 for compound **29**; however, compound **28** did not indicate this fragmentation pathway.

The authors draw attention to the difference in fragmentation of ion *m/z* 81 for compounds **28** and **29**. In compound **28**, ions *g*, *m/z* 81; *h*, *m/z* 80; and *l*, *m/z* 79 lose -HCN to generate the fragments *k*, *m/z* 54; *i*, *m/z* 53; and *m/z* 52, respectively (see **Figure 17**), at similar abundances, while **29** produces an abundant ion at *m/z* 52. The fact that the *m/z* 80 ion in **29** (*g*, **Figure 18**) releases N_2, but the same ion in **28** (*h*, **Figure 17**) does not, is related to the expansion of the ring structure for the formation of a pyrimidine fragment at **60** and a pyridazine at **29**. Compared to its isomers **28** and **29**, compound **30** has a low abundance of the molecular ion *m/z* 127. The fragmentation pattern of compound **30** (see **Figure 19**) indicates that the nitro group is absent; however, the presence of two adjacent groups gives rise to an *ortho* effect, which is responsible for two breaking patterns in this compound. The first is

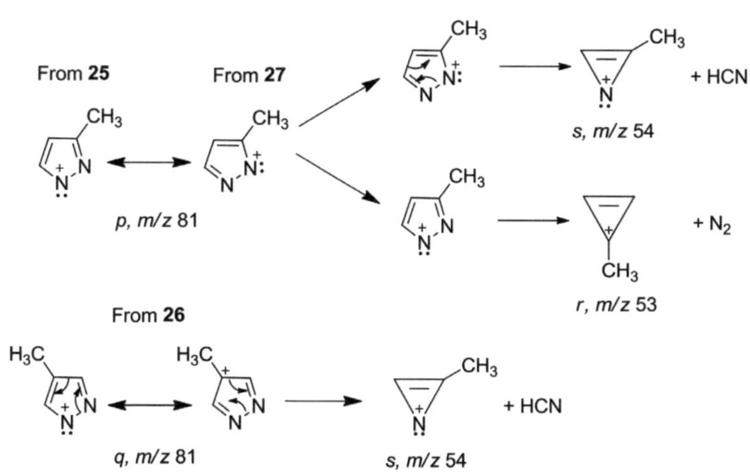

Figure 16.
Fragmentation of the $[M\ NO_2]^+$ of methyl-1-nitropyrazoles 25–27.

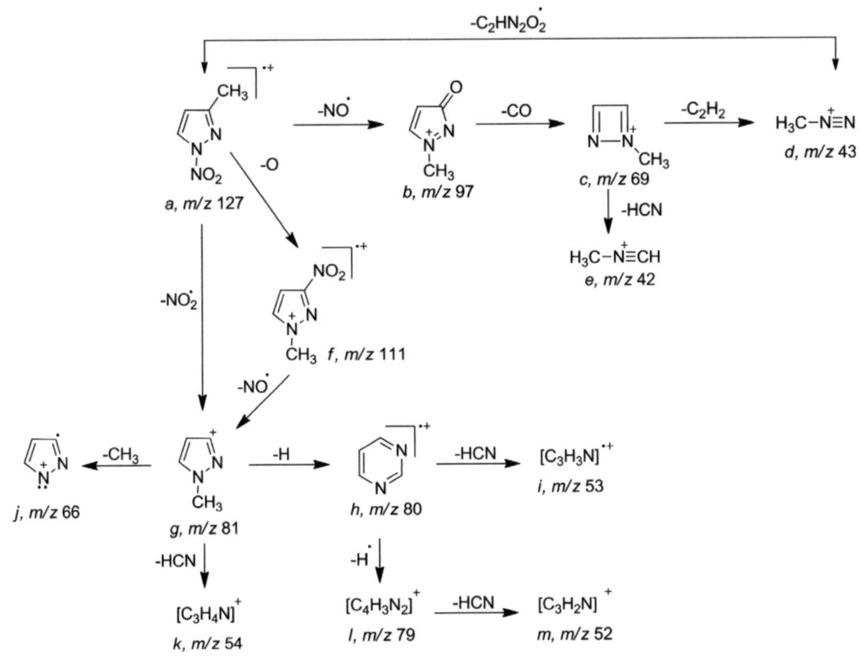

Figure 17.
*Fragmentation of 1-methyl-3-nitropyrazole **28**.*

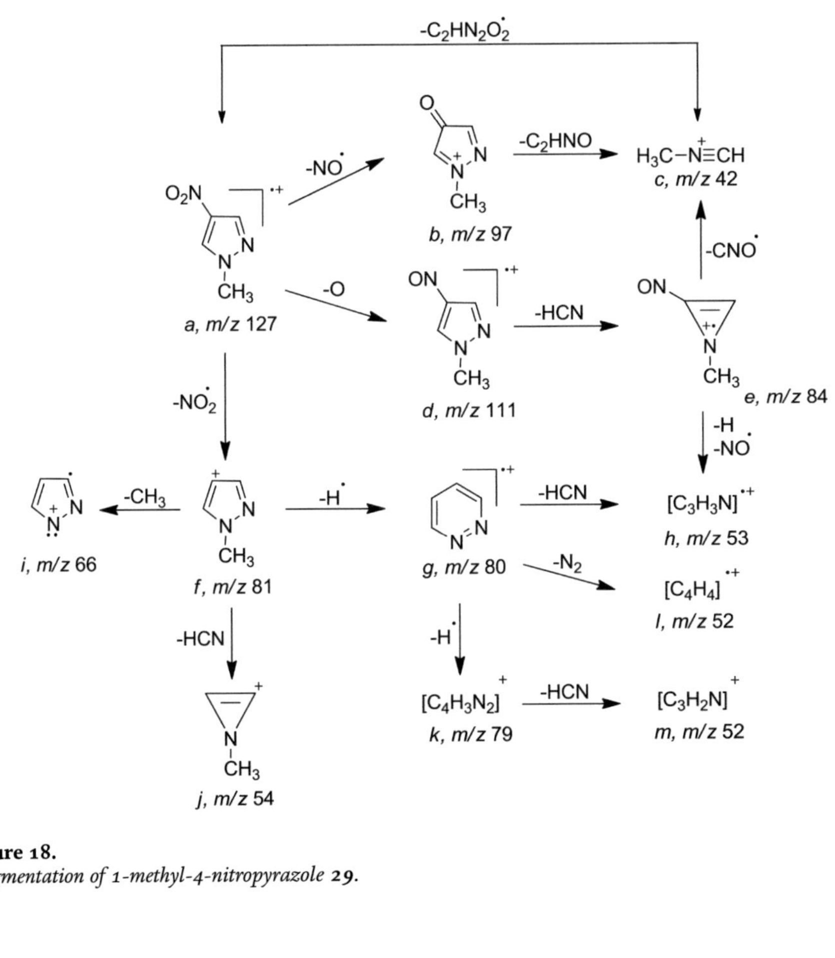

Figure 18.
*Fragmentation of 1-methyl-4-nitropyrazole **29**.*

the loss of OH• (*j*, *m/z* 110) from the molecular ion, with subsequent losses of NO• (*k*, *m/z* 80) and HCN (*l*, *m/z* 53). The $^{13}CH_3$ indicated that no ring expansion occurs and that the remaining oxygen atom—[M–OH]$^+$ ion, *m/z* 110—occupies its original position [5]. The second breaking pattern is less common and involves the loss of CHO• from the molecular ion, leading to the formation of *b*, *m/z* 98 (**Figure 19**). In a previous study, the authors showed that the carbon and hydrogen atoms that were lost originate from the methyl group [5]. The fragmentation of ion *b*, *m/z* 98 leads to the formation of ion *e*, *m/z* 68, which, according to the authors, must be identical to the pyrazole's molecular ion, since its subsequent fragmentation is analogous. Ion *b*, *m/z* 98 also shows a fragmentation path in which, after two successive losses of hydrogen radicals, it produces ion *g*, *m/z* 96, which can still suffer two losses—one forming ion *h*, *m/z* 68, through the loss of N_2, and the other forming ion *i*, *m/z* 66, through the loss of NO•. Unlike its isomers **28** and **29**, compound **30** does not show fragmentation of the molecular ion *a*, *m/z* 127, which leads to the formation of the *m/z* 42 or *m/z* 43 ions.

The mass spectrum of 3(5)-methyl-5-(3)-nitropyrazole (**31**) indicated a stable molecular ion—ions *g*, *m/z* 111; *e*, *m/z* 97; and *h*, *m/z* 81 confirmed the presence of the nitro group (**Figure 20**). One of the fragmentation pathways of the molecular ion *a*, *m/z* 127 generates ion *b*, *m/z* 110 through the release of •OH. Ion *b*, *m/z* 110 can still fragment in two ways: through the release of N_2 to form ion *c*, *m/z* 82, which, when it loses NO•, forms the ion *d*, *m/z* 52, and through losing NO• to form ion *k*, *m/z* 80. It is important to note that, similar to compounds **28** and **29**, ion *f*, *m/z* 41 is formed directly from the molecular ion by the loss of $C_2H_2N_2O_2$. Due to the contiguous substituents in compound **32** (**Figure 21**), two *ortho* effects are observed: (*i*) a loss of •OH, leading to the formation of ion *e*, *m/z* 110, which, after losing HCN, leads to ion *f*, *m/z* 83, and (*ii*) the loss of a H_2O, which gives a fragment of *m/z* 109, which can lose N_2 or HCN fragments. Other direct fragmentations of the molecular ion are the losses of NO_2• and NO•. As for pyrazole **28**, the loss of CO• from the *m/z* 97 ion to form ion *j*, *m/z* 69 is observed for compound **32**. This suggests that it is different from the *m/z* 69 ion of compound **28** (**Figure 17**), because for pyrazole **32**, the formation of the [CH_3–C≡N$^+$H] of *m/z* 42 through the loss of HCN is different from the *m/z* 42 ion formed in **28** ([CH_3–N$^+$≡CH]). The mass spectrum of pyrazole **32** has a second fragmentation at *m/z* 69 (*h* ion)—a possible route is the loss of the ion of H• and HCN from the [M–NO]$^+$ ion, *m/z* 97.

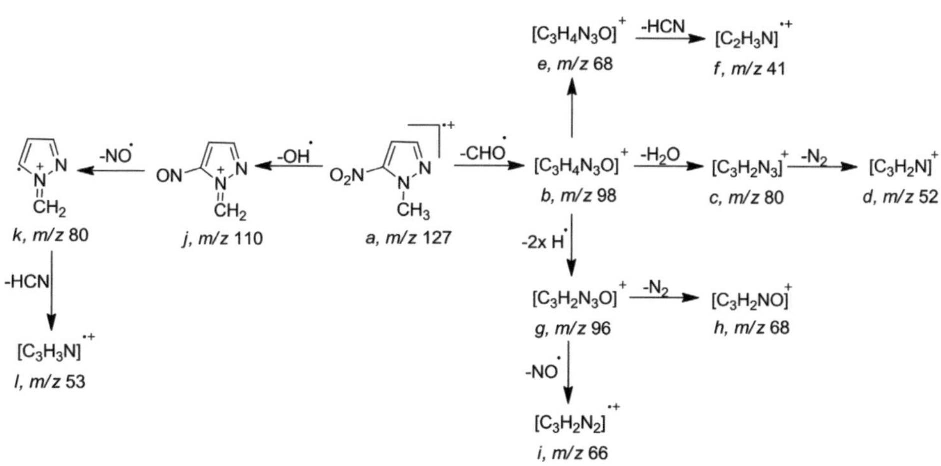

Figure 19.
Fragmentation of 1-methyl-5-nitropyrazole 30.

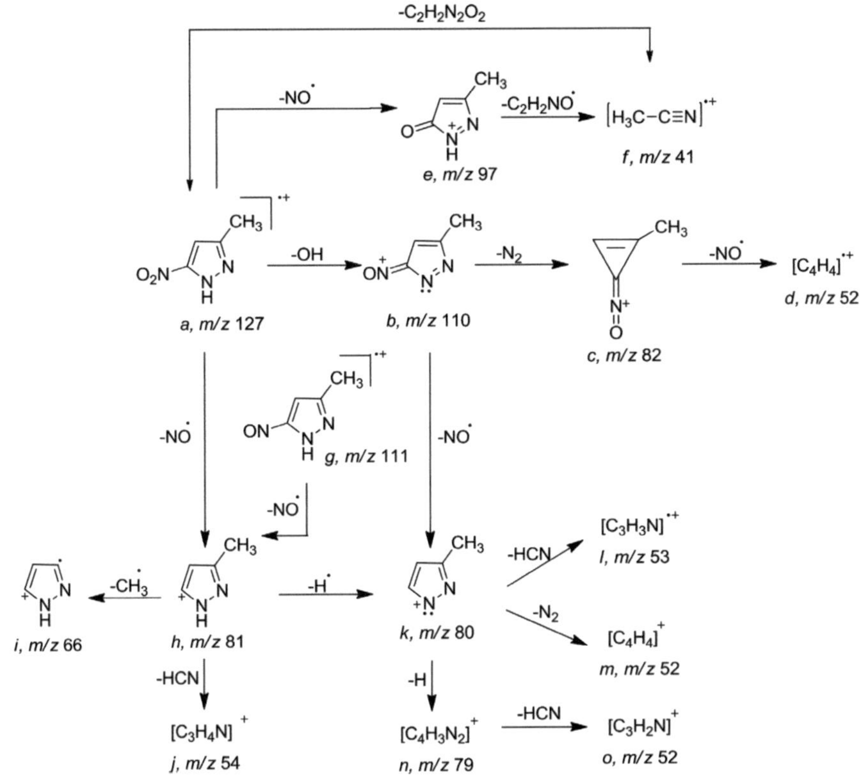

Figure 20.
Fragmentation of 3(5)-methyl-5(3)-nitropyrazole **31**.

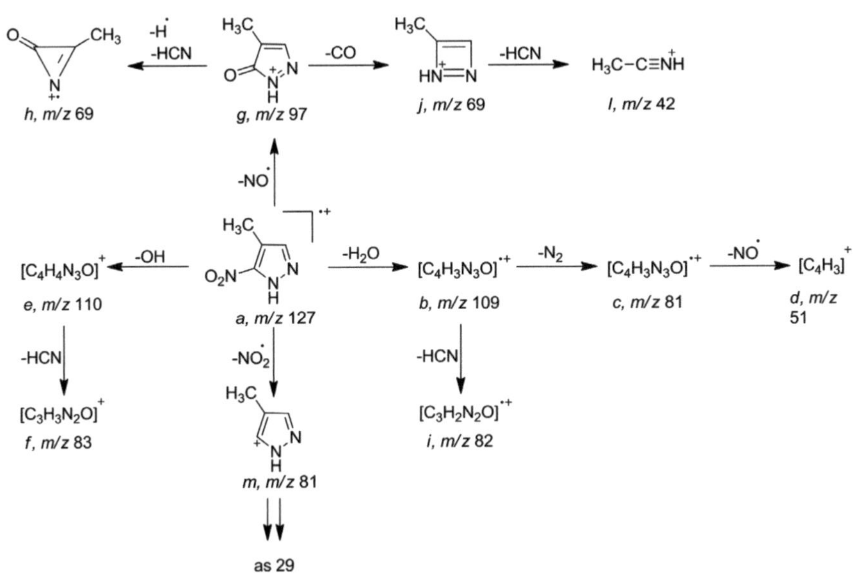

Figure 21.
Fragmentation of 3(5)-nitro-4-methylpyrazole **32**.

Similar to its isomers (**30** and **31**), pyrazole **33**—see **Figure 22**—had breaks due to the *ortho* effect: the primary losses of OH, CHO, and CH_2O. Ion *b*, *m/z* 110—which is formed by the loss of OH—is capable of releasing CO, thus forming ion *c*, *m/z* 82, which, through a loss of HCN, leads to the formation of ion *d*, *m/z* 55. Another interesting primary loss is $NO^{\bullet}$, which leads to the formation of ion *j*, *m/z*

97. Ion *j* is capable of releasing CO and, following the loss of a hydrogen radical, ion l, *m/z* 68 is formed. The predominant breakdown process ($[M–NO_2]^+$) to form *m*, *m/z* 81 followed the loss of a hydrogen in **31**, **32**, and **33**; therefore, the formation of the pyridazine molecular ion as in **29** would be expected. In all three cases (**31–33**), the product ion *m/z* 80 indicates the fragmentation behavior of this species: the loss of $H^•$, HCN, and N_2. Two other primary losses are those of $CHO^•$ and CH_2O. The $[M–CHO]^+$ ion, *i*, *m/z* 98, loses a hydrogen radical, followed by expulsion of HCN, to yield $[C_2H_2N_2O]^{+•}$, *h*, *m/z* 70, in contrast to the $[M–CHO]^+$ ion of **30**, which has an entirely different breakdown pattern (**Figure 19**).

The first fragmentation for pyrazoles substituted with NO_2 shows the rings intact, whereas the methyl substitution leads to the loss of a hydrogen radical with ring expansion. It is interesting to note that pyrazoles with $R^1 = NO_2$ (**25**, **26**, and **27**) lose this grouping in the first fragmentation, whereas the pyrazoles with $R^1 = CH_3$ do not suffer fragmentation of this group in the first breaks.

Tilborg and Thuijl [6] studied compounds **34–36**, **37–43**, and **44–47** (**Table 6**) using collision-induced dissociation (CID) mass spectrometry at low resolution, in order to understand how fragmentation occurs when pyrazoles are substituted at the 4-position. There are large differences between the ions derived from **34** and

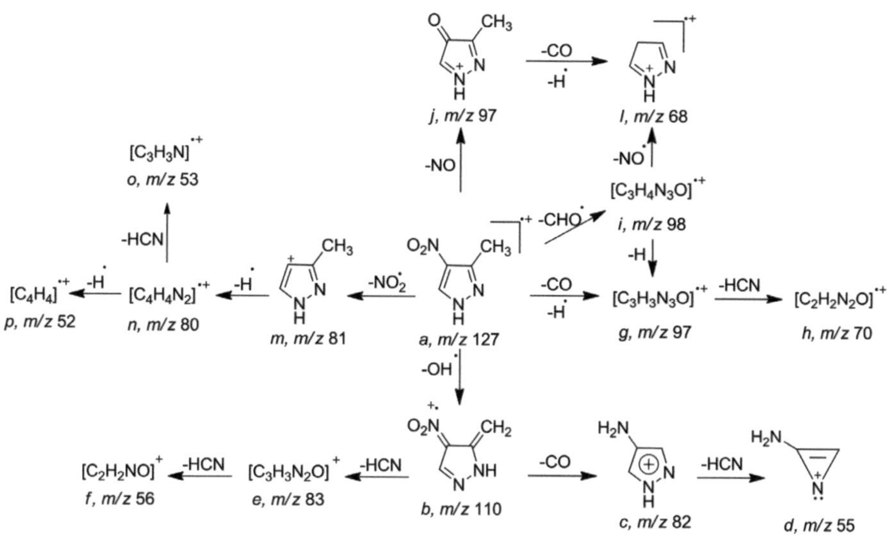

Figure 22.
Fragmentation of 3(5)-methyl-4-nitropyrazole 33.

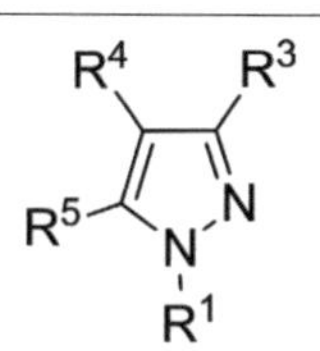

Comp.	R^1	R^3	R^4	R^5	Comp.	R^1	R^3	R^4	R^5
34	NO_2	H	H	H	42	D	Br	D	H
35	H	Br	H	H	43	D	Br	D	D
36	H	H	Br	H	44	D	H	Br	H
37	H	Br	D	H	45	H	D	Br	H
38	H	Br	H	D	46	H	D	Br	D
39	D	Br	H	H	47	D	D	Br	H
40	D	Br	H	D	48	D	D	Br	D
41	H	Br	D	D					

Table 6.
Structure of pyrazoles 32–48.

those from **35** and **36**. In 1-nitropyrazole **34**, the comparatively low abundance for hydrogen loss might be caused by the absence of hydrogen on the nitrogen rings.

The authors used the CID's corrected spectra of the deuterium analogs **35** and **36**, and they suggested that these—in addition to the **43** and **48** spectra—can be used to correct the spectra of the other compounds for the multiple losses of H (D). For each spectrum, the intensities were normalized for a total abundance of fragmentation equal to 100. From these data, approximate preference factors could be calculated for the collision-induced process $[C_3H_3N_2]^+ \rightarrow [C_3H_3N_2]^{+\bullet} + H^\bullet$. According to the authors, the losses are not specific to **35** or **36**, which suggests that a small amount of energy is required for the process to occur. The most abundant peak for the $[C_3H_3N_2]^+$ ion generated from **34** is *m/z* 39. The authors suggested that this may be due to $[C_3H_3]^+$ and/or $[C_2HN]^{+\bullet}$, formed by the loss of N_2 and/or HCN + $H^\bullet$, respectively. However, based on previous studies [11], it is indicated that $[C_3H_3]^+$ is more likely, due to the loss of N_2 being easier. These losses are compatible for the cyclic 1-pyrazolyl cation *a*, in which the positive charge is shared by all ring atoms **(Figure 23)**.

The differences between the correct CID spectra for $[C_3H_3N_2]^+$ ions generated from **37** and **38** were more difficult to understand. For these data, compounds **39–45** and **46–48** were labeled for further interpretation. These results allowed the authors to identify compositions of various fragment peaks. The ions in the regions between *m/z* 64 and *m/z* 68 do not provide any information on structural differences, but should be responsible for the loss of one or more hydrogens. From the behavior of the labeled compounds, the ions in the *m/z* 50–54 region must be formed by the loss of CH_2, $CH_3^\bullet$, $NH_2^\bullet$, and NH_3, respectively. The structurally significant differences in the *m/z* 67 ions for these compounds are in the *m/z* 36–42 region. The *m/z* 42 ion in **36** retains its marker when it is connected to the nitrogen atom. When the hydrogen's atoms attached to the carbons are labeled, one of them is retained, and this leads to the formation of *m/z* 44 at **40**, whereas in **48** this is absent. Thus, it is possible to conclude that the 4-bromopyrazole ion (**36**) can lose $C_2H^\bullet$. The *m/z* 41 and 40 fragments in **35** and **36** are due to the losses of $CN^\bullet$ and HCN, respectively. In the fragmentation of **35** and **36**, there is a predominant loss of $CH_2N^\bullet$. For compound **36**, the ion $[C_3H_3N_2]^+$ is five times more abundant than when generated from *m/z* 67 in **34** and **35**. This ion only retains its marker when it is attached at the nitrogen—it is not found if *m/z* is greater than 30. Thus, the authors assumed it should be $[NH_2]^+$. The *m/z* 26 and the *m/z* 28–24 fragments were exceptions and have similar abundances in **35** and **36**: *m/z* 28 $[N_2]^{+\bullet}$, *m/z* 27 $[HCN]^{+\bullet}$, *m/z* 25 $[C_2H]^{+\bullet}$, and *m/z* 24 $[C_2]^{+\bullet}$. Thus, the structure of the *m/z* 67 from **35** and **36** can be described as shown in **Figure 24**.

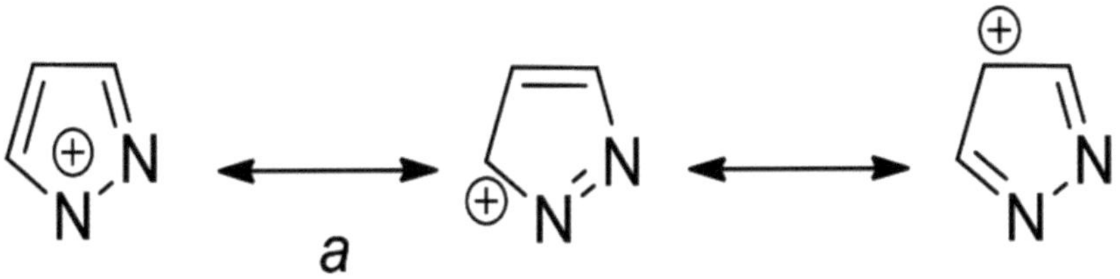

Figure 23.
Resonance structures from pyrazole ion.

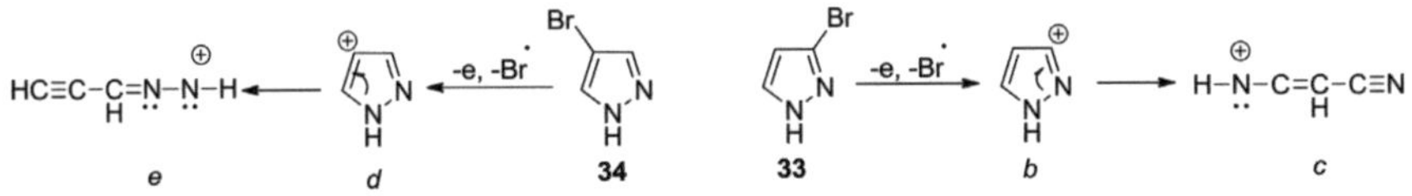

Figure 24.
Linear structure formed from fragmentation of 3(5)bromopyrazole.

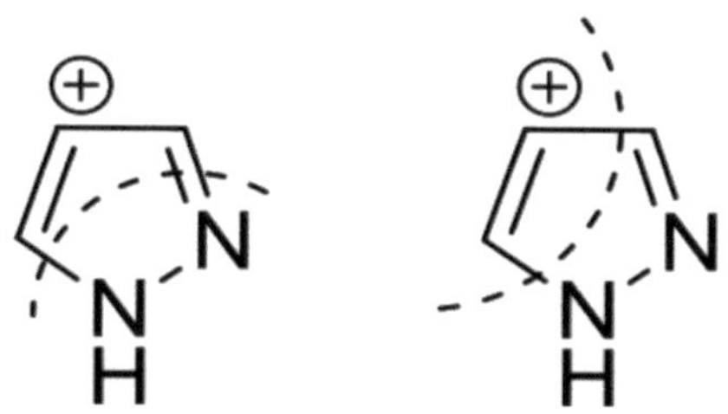

Figure 25.
Two possible cross sections for compound 34, based on the d *ion.*

The authors suggested that the peaks between *m/z* 41 and *m/z* 38 are formed due to the losses of CN•, HCN, and $CH_2N^{\bullet}$ and that the generation of $[C_2N]^+$ can be understood from ions *b*, *c*, and *d*; however, it cannot be formed from *e*. Therefore, the majority of $[C_3H_3N_2]^+$ ions formed from **36** have the cyclic structure of *d* rather than the linear structure of *e*. In contrast, most ions formed from **33** have structure *c*. The abundances of the fragments $[C_3]^{+\bullet}$, *m/z* 36 and $[C_3H]^{+\bullet}$, *m/z* 37 decrease in accordance with the following order: **34** > **35** > **36**. The formation of these ions can be explained by ions *a*, *b*, and *c* (**Figure 2**). The ions formed from **34** have the cyclic structure of *a*, whereas those generated from **36** mostly have the structure of *d*. Thus, the authors concluded that only a tiny fraction of the ions formed from **35** are cyclic, which is consistent with the presence of fragments *m/z* 29, $[N_2H]^+$ and *m/z* 42 $[CH_2N_2]^{+\bullet}$, in the corrected CID spectrum of $[C_3H_3N_2]^+$ ions generated from **36**. These fragments are readily formed from *d* (according to **Figure 25**) or *b* (**Figure 24**), but not *c* (**Figure 24**). The level of ion loss can be used advantageously to distinguish isomeric ions from different geometries (e.g., between cyclic and linear species). To achieve a given attenuation of the main beam, more target gas is needed for an apparently smaller cyclic ion than for a linear species.

3. Conclusions

In conclusion, unsubstituted, deuterated 4-chlorine, 4-bromine, phenyl, and methyl pyrazoles have stable molecular ions that generally form the base peak. The tandem fragmentation of these pyrazoles showed two important processes: (i) expulsion of HCN from $[M]^{+\bullet}$ and $[M\text{–}H]^+$ and (ii) loss of N_2 from $[M\text{–}H]^+$—the former process is the most predominant. However, substitution influences the relative importance of these two processes. While the patterns remain essentially unchanged in 4-bromo(chloro)pyrazoles, the introduction of more bromine atoms makes the initial loss of HCN less important because of the competing loss of bromine radicals. Secondary loss of HCN remains important, however, as it suppresses the expulsion of N_2. The influence of the 4-nitro and 4-acetyl substituents is much more pronounced, because the substituents can be in more than one step, giving rise to the stable ions $[M\text{–}O]^{+\bullet}$ and $[M\text{–}NO_2]^+$ in the former and to $[M\text{–}CH_3]^+$ in the latter case. In these compounds, the loss of HCN, though still strong as a secondary step, is no longer a primary process. When methyl- and nitro- groups are *ortho*-positioned, the pyrazole fragmentation sequence is affected. On the other hand, the 4-cyano and 4-phenyl substituents promote the loss of HCN in such a manner that the ion formed by the loss of the substituent is practically absent. In 4-cyanopyrazole, the initial loss of HCN is very strong, giving an ion with *m/z* 66, while in 4-phenylpyrazole, two consecutive losses of HCN occur. The fragmentation of monodeuterated, dideuterated, and trideuterated pyrazoles was similar to unsubstituted pyrazole but showed that the $H^{\bullet}$ is lost from C–H bond cleavage in an α-position relative to a nitrogen atom. The fragmentation of deuterated 3,5-diphenylpyrazoles differs from other diphenylpyrazoles; however, two similarities were found with the fragmentation of 4,5-diphenylpyrazoles: (i) the loss of N_2D and N_2H and (ii) the reversible transfer of a phenyl hydrogen to nitrogen, which, in the case of 1,3-diphenyl, occurs prior to or during the initial elimination of N_2D and N_2H. The fragmentation of 1-phenylpyrazole-4-yl-oximes depends on the substituent. The aldoximepyrazole suffered the loss of the following: water, a hydroxyl radical, and aldoxime radicals. In turn, the ketoximes suffered the loss of alkyloxime via a McLafferty-type rearrangement. Benzyl oxime pyrazoles behave differently from other compounds. Besides the anticipated loss of oxygen and a hydroxyl radical from the molecular ion, the unexpected ejection of NO, HNO, and H_2NO occurs. It appears that the molecular ions of these oximes do not undergo the Beckmann rearrangement, but rather a rearrangement of the oximino-hydrogen atom, which has not yet been encountered in the mass spectra of oximes. Finally, it is important to note that unfortunately the experimental gas chromatographic conditions were not informed by most of authors. Only Luijten and van Thuijl [4] informed the purity of compounds and it was checked on a GCMS system, using a 5% SE-30 packed column.

Acknowledgements

The authors are thankful for the financial support from the National Council of Scientific and Technological Development (Conselho Nacional de Desenvolvimento Científico e Tecnológico—CNPq) and also from the Rio Grande do Sul State Foundation for Research Support (Fundação de Amparo à Pesquisa do Estado do Rio Grande do Sul—FAPERGS)—Grant No. 16/2551-0000477-3 and Number 17/2551-0000 944-4. The fellowships from CNPq (C.P.F., M.A.P.M., N.Z., H.G.B., and A.V.P.) and CAPES (B.L.K, K.M.W., and B.L.H.) are also acknowledged.

Conflict of interest

There are no conflicts to declare.

Author details

Clarissa Piccinin Frizzo*, Bruno Luís Hennemann, Bruna Luisa Kuhn, Keli Maiara Wust, Alisson Vasques Paz, Marcos Antonio Pinto Martins, Nilo Zanatta and Hélio Gauze Bonacorso
Department of Chemistry, Núcleo de Química de Heterociclos (NUQUIMHE), Federal University of Santa Maria, Santa Maria, Brazil

*Address all correspondence to: clarissa.frizzo@gmail.com

References

[1] Faria JV, Vegi PF, Miguita AGC, Dos Santos MS, Boechat N, Bernardino AMR. Recently reported biological activities of pyrazole compounds. Bioorganic and Medicinal Chemistry. 2017;**25**:5891-5903

[2] van Thuijl J, Klebe KJ, van Houte JJ. The mass spectra of some pyrazole compounds. Organic Mass Spectrometry. 1970;**3**(12):1549-1559. DOI: 10.1002/oms.1210031213

[3] van Thuijl J, Klebe KJ, Van Houte JJ. The mass spectra of some deuterated pyrazoles. Journal of Heterocyclic Chemistry. 1971;**8**(2):311-312. DOI: 10.1002/jhet.5570080227

[4] Luijten WCMM, van Thuijl J. Mass spectrometry of nitroazoles. 1—The mass spectra of methyl substituted nitropyrazoles. Journal of Mass Spectrometry. 1979;**14**(11):577-584. DOI: 10.1002/oms.1210141102

[5] van Thuijl J, Klebe KJ, Van Houte JJ. Preference factors and isotope effects in the loss of H. (D.) and HCN (DCN) from deuterated pyrazoles upon electron-impact. Journal of Mass Spectrometry. 1971;**5**(9):1101-1105. DOI: 10.1002/oms.1210050910

[6] van Tilborg MWEM, van Thuijl J. Structure determination of [C3H3N2]+ ions, formed from monosubstituted pyrazoles C3H3N2R, by collision-induced dissociation mass spectrometry. Journal of Mass Spectrometry. 1984; **19**(11):569-572. DOI: 10.1002/oms.1210191108

[7] Bowie JH, Cooks RG, Lawesson S, Schroll G. Electron impact studies. XII. Mass spectra of substituted imidazoles. Australian Journal of Chemistry. 1967; **20**:1613-1624. DOI: 10.1071/CH9671613

[8] Aldous GL, Bowie JH. Electron-impact studies XCV-skeletal rearrangement fragments in the mass spectra of 3,5-diphenylisoxazole and 3,5-diphenylpyrazole. Journal of Mass Spectrometry. 1975;**10**(1):64-68. DOI: 10.1002/oms.1210100110

[9] Finar IL, Millard BJ. High-resolution mass spectrometry. Part VI. A hydrogen rearrangement in the mass spectra of pyrazole oximes. Journal of the Chemical Society C. 1969;**0**(19):2497-2499. DOI: 10.1039/J39690002497

[10] Khmel'nitskii RA, Krasnoshchek AP, Polyakova AA, Grandberg II. Mass spectra and structure of organic compounds. XX. Mass spectra of pyrazole and methylpyrazoles. Zhurnal Organicheskoi Khimii. 1967;**3**(9): 1540-1546 CA 68, 34199r (1968)

[11] Grubb HM, Meyerson S. In: McLafferty FW, editor. Mass Spectrometry of Organic Ions. New York: Academic Press; 1963. pp. 516-519 eBook ISBN: 9780323142779

Chapter 5

Gas Chromatographic-Mass Spectrometric Detection of Pesticide Residues in Grapes

Mahadev C. Khetagoudar, Mahadev B. Chetti and Dinesh C. Bilehal

Abstract

GC-MS/MS method has been developed and validated for the determination and quantification of 35 multi-class pesticide residues in grape samples. Pesticides are selected from different families including organochlorines, organophosphorus, carbamates, pyrethroids, triazines, triazoles, pyrazoles, etc. The QuEChERS-dSPE (dispersive solid-phase extraction) method was used for the extraction of residues of pesticide. An extra cleanup step was included with the help of a primary secondary amine (PSA) and graphitized carbon black (GCB). Recoveries ranged from 70 to 100% with 14% relative standard deviation (RSD). Other parameters such as precision, recoveries, limit of detection (LOD), limit of quantification (LOQ), and linearity were also studied. Finally, the proposed analytical method was successfully employed for the determination of residues of pesticide in grape samples.

Keywords: residues of pesticide, QuEChERS-dSPE, GC-MS/MS

1. Introduction

In India, a large quantity of pesticides is used for the cultivation of grapes mainly for the management of various diseases and pests. Due to the stringent rules set by the various developed countries on food safety standards and the regulations on quality parameters, we find that the residues of the pesticides in food are gaining a lot of attention. Keeping in view the problem of residues of pesticides, the present study was conducted on grape (*Vitis vinifera* L.) of Bijapur District for the qualitative and quantitative analysis of pesticides by GC-MS/MS (gas chromatography coupled to mass spectrometry).

In recent years, the production and marketing of food have gained topmost priority. This in turn has given rise for the implementation of better agricultural practices and has also prompted a substantial increase in the importance given to pesticide residues and related aspects. It is important to analyze large numbers of samples for residues of pesticide in the food due to their control and regulatory issues. Analytical procedures for pesticide residues are usually time-consuming and costly. For this reason multiresidue methods have been devised and regularly applied in regulating pesticide monitoring programmes [1, 2].

There is a difficulty in developing a method for the residue analysis mainly due to wider nature of polarity, volatility and solubility of different pesticides [3].

In relation with different pesticide classes, various methodologies using gas chromatography with numerous detectors, like thermal conductivity detector (TCD), nitrogen-phosphorus detector (NPD), electron capture detector (ECD) and flame photometric detector (FPD), have been implemented [4]. Further several methods have been developed for accurate quantification of residues of pesticides in various consumable food products or commodities. All these seem to be much complicated because of the use of large quantity of inert gases which are quite costly and consuming [5, 6]. Therefore, there is a need to develop new methods in the preparation of the sample and the requisite quantification parameters.

QuEChERS which is a novel quick, easy, cheap, effective, rugged, and safe method for preparation of samples in pesticide residue analysis [7] was used. QuEChERS methodology has been devised in the year 2003 for the multiresidue analysis of pesticides in different matrices, and now it is a universally accepted method. In this procedure extraction was performed with acetonitrile solvent initially and then partitioning step was carried out using salt mixture. A small amount of extract was further cleaned by using dispersive solid-phase extraction (dSPE) method. Finally, extract was used for the determination of pesticide residues using GC-MS/MS. This method has several advantages; firstly, sample throughput is very high; secondly, it does not use chlorinated solvents; and thirdly, a very small quantity of solvents is needed which in turn provides a very high recovery percentage for broad-spectrum volatility and polarity range of pesticide molecules. Even though this method was developed recently, it has been widely accepted by the international community of pesticide residue analysts. There have been several publications on this topic often replacing the original method with newer and better ones [8–12].

Chromatographic system (gas chromatography or liquid chromatography) attached to mass spectrometry (MS/MS) determination provides us with a method for identifying and quantifying several pesticides in different food matrices [13]. Simple extraction procedure along with very limited cleanup technologies has been employed as a result of the use of more sensitive and selective MS/MS detection. Martinez Vidal et al. used gas chromatography-mass spectroscopy (GC-MS/MS) with ethyl acetate for extraction of 130 multi-class pesticides [14]. Pihlström et al. slightly modified GC-MS/MS procedure [15]. Hetherton et al. reported the use of LC-MS/MS and acetonitrile extraction for the analysis of 73 pesticides in lettuce and oranges [16]. Pang et al. used both liquid chromatography and gas chromatography attached to mass spectrometry for the simultaneous determination of 336 pesticides in vegetables and fruits [17, 18] and 440 pesticide residues in wine, fruit juice, and honey using solid-phase extraction (SPE) cleanup [7].

Grape (*Vitis vinifera* L.) is one of the most important fruit crops cultivated in the subtropical regions of India (60,000 ha). The states, namely Maharashtra, Karnataka, Andhra Pradesh, Tamil Nadu, Punjab, and Haryana, are the grape-growing regions in India. Amongst them, Maharashtra and Karnataka rank first and second in terms of area and productivity, respectively. Grapes cultivated in Maharashtra and Karnataka are mainly exported to Europe, the Middle East and to some extent West Asia. As a result, a large quantity of pesticides is used in their cultivation. This is mainly due to the presence of heavy insect pest infestation. Excess usage of pesticides often results in the accumulation of pesticides on the fruit and causes various health hazards and is also more prone for rejection in the international market.

This paper explains an effective and simple experimental procedure for extraction of sample by employing QuEChERS (slightly modified) method and the use of gas chromatographic system with mass spectrometric determination for 35 pesticide residues in grape samples.

2. Experimental

2.1 Apparatus

a. GC-MS/MS instrument: gas chromatograph (Agilent 6890N) with auto-sampler and a triple quadrupole mass spectrometer (Quattro Micro RAB120 Waters) detector was used for the analysis of the pesticides studied. MassLynx Solution software was used for the instrument control and data analysis.

b. Low-volume concentrator: Turbovap (Caliper Life Sciences, USA) with inert nitrogen was used for the evaporation of the solvent.

c. Chopper and homogenizer: vegetable chopper was used for chopping, and a homogenizer (Heidolph) was used for proper mixing of the fruit samples.

d. Centrifuge: centrifuge (Sigma 3K 10) was used for both 2 and 50 ml polypropylene tubes.

e. Weighing balance: weighing balance (Sartorius) was used to weigh the chopped samples and preparation of reference standard reagents.

2.2 Reagents

a. Ethyl acetate and acetic acid (glacial): ethyl acetate and acetic acid (glacial) of sufficient quality for pesticide residue analysis were procured from Sigma-Aldrich.

b. Sodium acetate and magnesium sulfate: reagent-grade anhydrous sodium acetate and magnesium sulfate were procured from Merck (India).

c. Certified reference materials (CRMs): certified reference materials of pesticides were procured from Sigma-Aldrich/Riedel-de-Haen (Zwijndrecht, The Netherlands). The individual stock solutions of 1000 ppm were prepared in toluene and hexane (1:1), and working standards containing 35 pesticides at different concentration levels were prepared in ethyl acetate.

d. Primary secondary amine (PSA): SPE sorbent PSA (40 μm, Bondesil PSA) was purchased from Agilent Technologies (Bangalore, India).

e. Grape samples: grape samples (2 kg each) were collected from the field in Vijayapura district (Karnataka state).

2.3 Residue extraction and cleanup step

The method of preparation of the sample for multiresidue pesticide analysis in grapes involved the following steps: (1) crush 2 kg grape samples under ambient conditions and then 200 g of sample further homogenized for 2 min for proper mixing; (2) accurately weigh a 10 ± 0.1 g of this sample into each 50 ml

Sl. no.	Reference standards	t_M (min)	MRM		CE	Fortification levels (mg/kg)		LOD (mg/kg)
			Pre. ion	Prod. ion		0.02	0.05	
1	DEET	7.06	119	65	21	90 (7)	95 (12)	0.001
2	Propiconazole	7.65	69	41	6	88 (14)	92 (11)	0.01
3	Phorate	7.85	260	75	5	69 (9)	75 (12)	0.002
4	Carbofuran	8.35	164	149	8	78 (4)	84 (7)	0.002
5	Atrazine	8.45	215	58	8	95 (7)	99 (5)	0.005
6	Lindane	9.16	184	145	10	90 (5)	95 (8)	0.001
7	Diazinon	9.74	179	137	17	79 (16)	87 (12)	0.0005
8	Chlorothalonil	9.95	266	133	26	84 (16)	86 (9)	0.004
9	Metalaxyl	10.37	206	59	8	74 (9)	76 (13)	0.002
10	Fenitrothion	10.64	125	79	11	89 (9)	91 (4)	0.002
11	Malathion	10.70	173	99	10	91 (7)	89 (11)	0.003
12	Aldrin	11.54	263	193	22	91 (7)	89 (11)	0.003
13	Fenthion	11.99	278	109	12	80 (11)	92 (15)	0.005
14	Chlorpyrifos	12.05	197	169	16	93 (8)	96 (7)	0.0005
15	Parathion	12.39	291	109	10	83 (6)	95 (12)	0.003
16	Triadimefon	12.77	208	181	6	88 (9)	91 (10)	0.006
17	Pendimethalin	13.39	252	162	16	95 (3)	98 (13)	0.005
18	Captan	13.95	79	51	20	72 (10)	79 (8)	0.002
19	Phenthoate	14.19	274	121	16	88 (4)	90 (6)	0.0005
20	2,4-DDT	14.61	146	118	7	70 (15)	82 (10)	0.00001
21	Alpha-endosulfan	14.95	241	170	25	70 (12)	82 (9)	0.004
22	Butachlor	15.29	176	146	20	94 (8)	99 (13)	0.001
23	Profenofos	15.76	337	267	8	73 (11)	78 (6)	0.005
24	2,4-DDD	16.34	235	165	16	95 (11)	98 (9)	0.00001
25	Endrin	16.85	263	193	22	83 (6)	86 (5)	0.005
26	Chlorfenapyr	17.15	247	75	17	76 (5)	78 (13)	0.02
27	Beta-endosulfan	17.41	241	170	25	70 (14)	87 (9)	0.005
28	Quinalphos	17.87	235	165	15	98 (9)	100 (5)	0.003
29	Ethion	17.89	231	129	18	91 (14)	96 (10)	0.0001
30	Triazophos	18.72	161	77	19	72 (5)	79 (10)	0.005
31	Iprodione	18.91	314	245	10	87 (7)	90 (4)	0.02
32	Beta-cyfluthrin	19.63	165	127	5	96 (15)	99 (8)	0.01
33	Alpha-cypermethrin	20.17	163	127	6	91 (11)	95 (8)	0.005
34	Fenvalerate	20.72	167	125	8	89 (9)	94 (3)	0.005
35	Deltamethrin	21.73	181	152	18	70 (15)	78 (11)	0.008

t_M, retention time; MRM, multiple reaction monitoring; CE, collision energy; LOD, limit of detection; LOQ, limit of quantification.

Table 1.
Experimental condition of the optimized GC-MS/MS method parameters, retention time (min), MRM, average recovery (%) and RSD (in parenthesis) of grape samples (n = 4) at two concentration levels.

polypropylene tubes; (3) add 10 ± 0.1 ml acetonitrile (1% acetic acid) to the polypropylene tubes; (4) homogenize the sample at 3000–5000 rpm for 2–3 min. Add 1.5 g sodium acetate and 6 g $MgSO_4$ (anhydrous), and mix it by shaking gently and centrifuging at 3000 rpm for 3 min to separate the organic layer; (5) 1 ml of extract is then taken in a separate dSPE (dispersive solid-phase extraction) tube, and 50 mg PSA and 140 mg magnesium sulfate are then added to it; (6) extracts were then centrifuged at 5000 rpm for 1 min; (7) 1 ml of the supernatant was then transferred to a small glass tube and the solvent was then evaporated using turbo evaporator which was set at 45°C and 20 psi inert nitrogen gas flow; and (8) the final step was the reconstitution of the sample with a 1 ml of ethyl acetate for analysis and confirmation of residues by gas chromatography-mass spectrometry (GC-MS/MS) (24).

GC-MS/MS analysis. A gas chromatography (Agilent 6890N) with mass spectrometer (Waters, Boston, USA) and an auto-sampler (Agilent 7683) with electron ionization (EI+) mode were used. Separation of analytes was carried out using HP-5MS column (30 m, 0.25 μm internal diameter, 0.25 μm film thickness) (J&W Scientific). The oven temperature was increased as follows: 50°C (1 min), 25°C/min up to 150°C and increased to 10°C/min to 280°C (4 min hold). Split-less injection of 1 μl was carried out with an injector temperature maintained at 280°C and hold time of 1 min. The carrier gas that was used was helium (99.999%) at flow rate of 1.3 ml/min. The interface temperature was maintained at 250°C.

Electron ionization (EI+) mode was selected (4 min solvent delay); the source temperature was set at 250°C. The gas argon (purity 99.99%) was used as collision gas which is used to collide with ions after ionization. The dwell time per channel was between 0.05 and 0.1 s. QuantLynx was used to process the data obtained from calibration of CRMs and also from grape fruit extract.

Heptacosa (perfluorotributylamine) was used to calibrate the mass spectrometer. **Table 1** explains the particular ions of quantification for the MRM mode and retention times (t_M) for the residue analysis of individual substances.

3. Validation study

In this method, for the fulfillment of validation criterion, single-laboratory approach was used. The following validation parameters were used.

3.1 Linearity

Five calibration levels (1 and 200 ng/mL) were used for constructing the calibration curve by using pure solvent and matrix. The concentration of a pesticide residue can be calculated based upon the calibration curve. The prerequisite for this method is that the peak area should fall within the linear range of the curve. Then the concentration can be calculated on basis of the slope of the calibration curve using the regression equation:

$$Y = mX + C \tag{1}$$

where Y = peak area, X = concentration, m = slope of the curve and C = constant.

3.2 Selectivity

It was determined by elimination of noise at the retention time of the compound, which is performed by fixing two transitions of MS/MS for individual molecule of analyte by considering the adequate precursor and product ions.

3.3 Sensitivity

Detection limit (LOD) in the chromatogram was calculated by using peak signal of the analyte molecule concentration to the three times background noise in the chromatogram. The quantification limit (LOQ) in the chromatogram was set as the lowest concentration with very good recovery range (65–100%) and precision (RSD ≤ 20%). The ion ratio (Q/q) was used for the criterion of confirmation in positive samples. The Q/q is the ratio of the intensity quantification (Q) and confirmation transition (q) (**Table 1**).

Typical S/N acceptance criteria: LOD—3:1 and LOQ—10:1.

4. Results and discussion

4.1 Method validation

All the 35 certified reference materials of pesticides were determined in a single chromatographic window of 22 min run time (**Figure 1**). For most of the compounds, the R^2 values (correlation coefficients) of the calibration curve were >0.99 for both pure solvent-based and matrix-matched samples. The recovery of all of the compounds was found to be 70–100% with RSD below 14%. **Figure 2** shows chromatograms of typical pesticide peak shapes. The grape sample extracts were slightly yellow because of the co-extraction of carotenoids in the ethyl acetate. It was found that a two-step homogenization procedure significantly increased the precision of analysis. It was also observed that the use of high-speed homogenization process

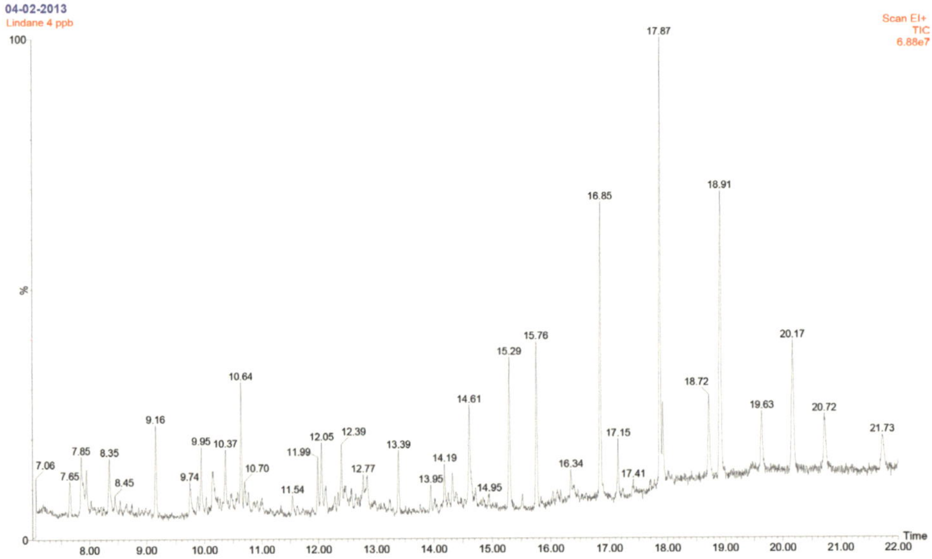

Figure 1.
Total ion chromatogram of 35 certified reference standards.

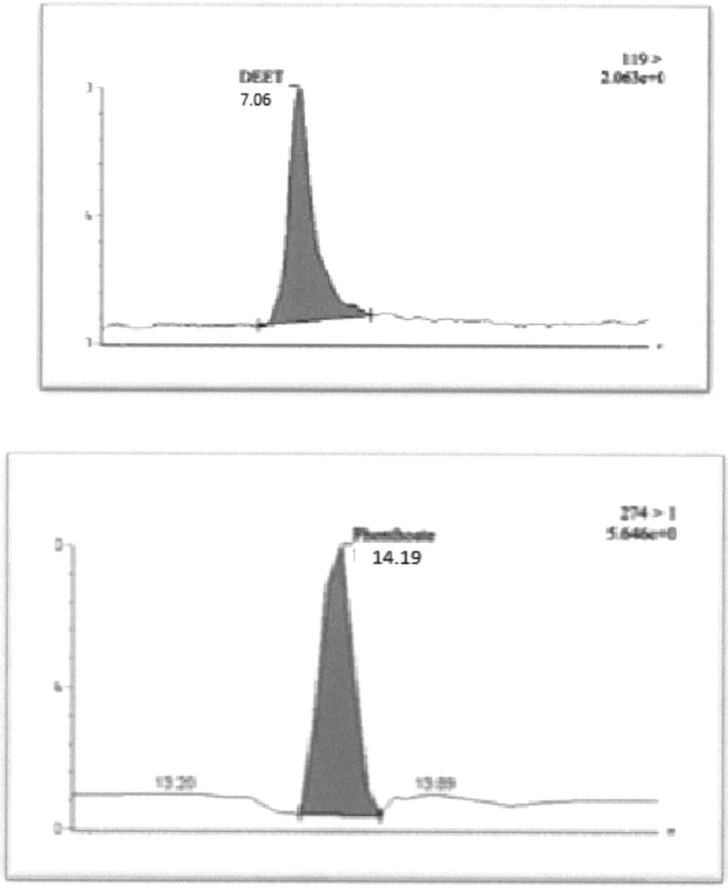

Figure 2.
Chromatograms showing DEET and phenthoate peak shapes.

for precooled sample did not necessarily increase the temperature of the system above 10°C and thereby proved very useful for maintaining the stability of the phthalimides such as captan.

4.2 Recovery experiments of spiked samples

Usually, the extraction and cleanup procedure removes the matrix co-extractives then separates all of the analytes from the matrix. The same does not holds good in most of the matrices during the pesticide residue analysis. As a result, the actual recovery experiments were performed on grape samples. The separated peaks with their t_M (retention times) are summarized in **Table 1**. Using the linear regression equation recoveries of individual pesticides with different levels of spiking along with replicates were calculated in grape matrix. **Table 1** gives the average recoveries for all spiked pesticide standards at each spiked level in grape samples. All the tested 35 pesticides displayed a recovery range between 70 and 100% which is quite acceptable. RSD (relative standard deviation) was used to express the reproducibility, and most of the RSD values were found to be less than 14%. Recovery study is conducted using a control sample and at least two fortification levels with three replications. The formula for arriving percentage recovery for method validation is as follows:

$$\text{Residue (ppm)} = \frac{\text{Area of sample}}{\text{Area of std}} \times \frac{\text{Final vol (ml)}}{\text{Vol injected } (\mu l)} \times \frac{\text{Std conc (ng)}}{\text{Wt of the sample(g)}} \times \frac{100}{\text{Fortification level}} \quad (2)$$

In QuEChERS method the use of acetonitrile has several advantages, mainly addition of salt separates it from water without using nonpolar solvents, mutual compatibility with dispersive solid-phase extraction and very good separation/matching with gas and liquid chromatography. The anhydrous $MgSO_4$ tends to form lumps. Shaking the tubes of the centrifuge on adding the salt mixture for 1 min or more, it was observed that the formation of lumps was eliminated. Next, adding the salt to all of the samples, it was also found that the one-minute extractions of the entire batch could be run parallelly. Dispersive solid-phase extraction with primary secondary amine eliminated the color pigments, acidic components and sugars [10, 11]. Apart from this, sugars, lipids and waxes were removed by freezing which helped in increasing the efficiency of GC analysis [12].

5. Applicability of the developed method

5.1 Sampling

Grape samples were collected from farmers' fields in Vijayapura district in Karnataka. These areas are very popular for the production of grapes, and we see an excessive use of pesticides. The developed analytical method was used for the determination of residues of pesticides in grape samples and that were analyzed in triplicate. The results confirmed that the grape samples contained pesticide residues well above the prescribed level, viz. carbofuran, fenvalerate, triazophos, and endrin (**Table 2**). Grapes which were analyzed in the present study mainly contributed to the major dietary intakes of the citizens in India.

Sl. no.	Name of the pesticides	MRLs exceeded samples	Residue content (ppm)	EU MRLs (ppm)
1	Carbofuran	9	0.14	0.02
2	Fenvalerate	6	0.33	0.02
3	Triazophos	4	0.13	0.01
4	Endrin	5	0.04	0.01

EU, the European Union; MRL, maximum residue limit.

Table 2.
Analytical results of grape sample analysis collected from Vijayapura district (n = 100).

6. Conclusion

Grapes contaminated with residues of pesticides pose a major health hazard. Therefore we have developed effective method for the detection of contaminated grapes. Hence, for the simultaneous confirmation and quantification of 35 pesticides in grape samples, a mulitresidue method has been developed and validated. For multi-class pesticide residue determination, GC-MS/MS with triple quadrupole analyzer played an important role. Within 22 min of run time, all the closely eluted and co-eluted peaks were separated with higher sensitivity. The two MRM transitions, one for confirmation another for quantification, achieved very good sensitivity and selectivity for possible safe identification by the use of Q/q ratio parameter. The limit of detection was lower than the MRL prescribed. Solid-phase extraction with acetonitrile solvent was employed. Finally, the method was successfully validated for two concentrations, viz. 0.02 and 0.05 mg/kg grape sample. The validated method reduces the overall cost of analysis and also offers low-uncertainty measurement. Further, this method was successfully employed for the analysis of real-world grape samples.

Acknowledgements

Authors acknowledge the ASIDE, Govt. of India and University of Agricultural Sciences, Dharwad, Karnataka, for providing grants for the establishment of Pesticide Residue Testing and Quality Analysis Laboratory.

Conflicts of interest

There are no conflicts of interest to declare.

Author details

Mahadev C. Khetagoudar[1], Mahadev B. Chetti[2] and Dinesh C. Bilehal[1*]

1 P.G. Department of Chemistry, Karnataka University, Dharwad, Karnataka, India

2 Indian Council of Agricultural Research, New Delhi, India

*Address all correspondence to: drbilehal@gmail.com

References

[1] McMahon BM, Hardin NF. Pesticide Analytical Manual. Vol. 1. Washington, DC: Food and Drug Administration; 1994, ch. 3

[2] Fillon J, Hindle R, Lacroix M, Selwyn J. Multiresidue determination of pesticides in fruits and vegetables by gas chromatography-mass selective detection and liquid chromatography with fluorescence detection. Journal of AOAC International. 1995;**78**(5):1252-1266

[3] Sivaperumal P, Anand P, Riddhi L. Rapid determination of pesticide residues in fruits and vegetables, using ultra-high-performance liquid chromatography/time-of-flight mass spectrometry. Food Chemistry. 2015;**168**:356-365

[4] Jardim ANO, Mello DC, Goes FCS, Junior EFF, Caldas ED. Pesticide residues in cashew apple, guava, kaki and peach: GC–μECD, GC–FPD and LC–MS/MS multiresidue method validation, analysis and cumulative acute risk assessment. Food Chemistry. 2014;**164**:195-204

[5] Albero B, Sánchez-Brunete C, Tadeo JL. Multiresidue determination of pesticides in juice by solid-phase extraction and gas chromatography–mass spectrometry. Talanta. 2005;**66**(4):917-924

[6] Štajnbaher D, Zupančič-Kralj L. Multiresidue method for determination of 90 pesticides in fresh fruits and vegetables using solid-phase extraction and gas chromatography-mass spectrometry. Journal of Chromatography A. 2003;**1015**(1-2):185-198

[7] Anastassiades M, Lehotay SJ, Štajnbaher D, Schenck FJ. Fast and easy multiresidue method employing acetonitrile extraction/partitioning and "dispersive solid-phase extraction" for the determination of pesticide residues in produce. Journal of AOAC International. 2003;**86**(2):412-431

[8] Lehotay SJ, Kok AD, Hiemstra M, Bodegraven PV. Validation of a fast and easy method for the determination of residues from 229 pesticides in fruits and vegetables using gas and liquid chromatography and mass spectrometric detection. Journal of AOAC International. 2005;**88**(2):595-614

[9] Diez C, Traag WA, Zommer P, Marinero P, Atienza J. Comparison of an acetonitrile extraction/partitioning and "dispersive solid-phase extraction" method with classical multi-residue methods for the extraction of herbicide residues in barley samples. Journal of Chromatography A. 2006;**1131**(1-2):11-23

[10] Lehotay SJ, Maštovská K, Yun SJ. Evaluation of two fast and easy methods for pesticide residue analysis in fatty food matrixes. Journal of AOAC International. 2005;**88**(2):630-638

[11] Lehotay SJ, Maštovská K, Lightfield AR. Use of buffering and other means to improve results of problematic pesticides in a fast and easy method for residue analysis of fruits and vegetables. Journal of AOAC International. 2005;**88**(2):615-629

[12] Vidal JM, Liébanas FA, Rodriguez MG, Frenich AG, Moreno JF. Validation of a gas chromatography/triple quadrupole mass spectrometry based method for the quantification of pesticides in food commodities. Rapid Communications in Mass Spectrometry. 2006;**20**(3):365-375

[13] Carneiro RP, Oliveira FA, Madureira FD, Silva G, de Souza WR, Lopes RP. Development and method validation for determination of 128 pesticides in

bananas by modified QuEChERS and UHPLC–MS/MS analysis. Food Control. 2013;**33**(2):413-423

[14] Martinez Vidal JL, Arrebola FJ, Garrido Frenich A, Martinez Fernich A, Martinez Fernandez J. Validation of gas chromatographic-tandem mass spectrometric method for analysis of pesticide residues in six food commodities. Selection of reference matrix for calibration. Chromatographia. 2004;**59**:321-327

[15] Pihlström T, Blomkvist G, Friman P, Pagard U, Österdahl BG. Analysis of pesticide residues in fruit and vegetables with ethyl acetate extraction using gas and liquid chromatography with tandem mass spectrometric detection. Analytical and Bioanalytical Chemistry. 2007;**389**(6):1773-1789

[16] Hetherton CL, Sykes MD, Fussell RJ, Goodall DM. A multi-residue screening method for the determination of 73 pesticides and metabolites in fruit and vegetables using high-performance liquid chromatography/tandem mass spectrometry. Rapid Communications in Mass Spectrometry. 2004;**18**(20):2443-2450

[17] Pang GF, Fan CL, Liu YM, Cao YZ, Zhang JJ, Li XM, et al. Determination of residues of 446 pesticides in fruits and vegetables by three-cartridge solid-phase extraction gas chromatography-mass spectrometry and liquid chromatography-tandem mass spectrometry. Journal of AOAC International. 2006;**89**(3):740-771

[18] Pang GF, Fan CL, Liu YM, Cao YZ, Zhang JJ, Fu BL, Li M, z-y, Wu YP. Multi-residue method for the determination of 450 pesticide residues in honey, fruit juice and wine by double-cartridge solid-phase extraction/gas chromatography-mass spectrometry and liquid chromatography-tandem mass spectrometry. Food Additives and Contaminants; 2006;**23**(8):777-810

Chapter 6

Gas Chromatography in Food Authentication

Kristian Pastor, Marijana Ačanski and Djura Vujić

Abstract

Authentication of food products and food fraud detection are of great importance in the modern society. The application of sophisticated instrumentation, such as gas chromatography (GC), with this aim helps to improve the protection of consumers. Gas chromatography mostly combined with the most powerful detector, a mass spectrometer (MS), and various multivariate data processing tools is in the last few decades being increasingly applied in authenticity and traceability of a wide spectra of food products. These include animal and plant products, beverages and honey. This chapter gives an overview of the most recent applications of gas chromatography technique in determining food authenticity, described in scientific literature.

Keywords: food products, authenticity, food fraud, consumer protection, gas chromatography

1. Introduction

The adulteration practices on food product market are known since ancient times [1, 2]. It was found that, during the nineteenth century, gypsum and alum were added to bakery flour to increase weight, strychnine was added to beer to increase bitterness, and salts of copper, lead, and mercury were added to sweets in order to get a beautiful color and gloss [3–5]. Consumer interest in safety, authenticity and quality of food products is constantly increasing [6]. Authenticity is related to truthfulness, so a food product can be said to be authentic if it was not subject to any fraud [7]. European and global food policies require food on the market to be authentic. This means that the label on the product must match its actual composition, origin (geographical, botanical and genetic) and the process of production (conventional, organic and traditional) [2, 8, 9]. With globalization, market development and rapid distribution systems, as well as expanding the range of food items, counterfeiting and contamination of food products, are becoming international in character, and the possible consequences are far-reaching [2, 4, 9–11]. The most common type of adulteration—economically motivated food adulteration—is defined as *a misleading and deliberate substitution or addition of certain ingredients to a food product in order to increase the apparent value of the product or reduce the cost of its production, with the consequence of a certain economic gain* [4, 5]. Depending on the nature of an added substituent, the obtained adulterated products may pose a potential danger to the health of the consumer. In this way, the determination of authenticity in the food industry is gaining health and safety aspects, in addition to the economic one [6, 8, 12]. With all this in mind, global policies require strict monitoring and quality control of food. Therefore, there is a clear tendency toward the development of new techniques and analytical methods that would enable this goal to be achieved. Traditional and standard methods of analysis are still very commonly used. Due to lower costs and/or faster analytical protocols, there is an urge for new authentication methodologies that would be complementary or even replace existing ones [8, 9]. This trend is stimulated by consumers, regulatory bodies and the food industry itself. Contemporary authentication analysis is based on the detection and measurement of various chemical parameters that would have the potential of discrimination factors of the investigated food samples [2, 9]. According to Danezis et al. [2], the first 10 countries in the world that are most intensively engaged with food authentication, in addition to the United States and China, are members of the European Union. These countries actively subsidize and encourage the development of this scientific area [2]. The European Commission regulations and directives testify about the rights of consumers to get the genuine information about food products that they buy [13–15]. These regulations aim to prevent (i) fraud and misleading actions, (ii) adulteration of food products and (iii) any other fraudulent procedures. An example of a very frequent way of food adulteration is the substitution of some ingredient in a food item with a similar and cheaper one, so that the consumer cannot recognize this procedure [1, 6, 8, 16]. According to the literature data, food products mostly subjected to adulterations include cereal and bakery products, edible oils and fats, milk and dairy products, meat and fish, fruit and fruit juices, honey, coffee, tea, wine, organic products and many others [9, 11]. Basically, there are three analytical approaches to determine the authenticity of food products: (i) chemical approach, determination of the composition and content of various chemical components in food; (ii) biomolecular approach, analysis of DNA and proteins; and (iii) isotopic approach, determination of the composition of stable isotopes of certain atoms [7]. Chromatographic techniques are the most common choice in the analysis of the authenticity of most food items [2, 9]. This is partly because techniques, such as chromatography, can be applied both for the purpose of detecting adulterations and for the purpose of determining authenticity [7]. In addition, the analytical capability of mass spectrometry, often used in conjunction with chromatographic techniques, allows the characterization of a wide range of components in very complex systems [17]. Some authors believe that the future of determining food authenticity is reflected in the synergistic fusion of various complementary instrumental techniques and the processing of such a complex block of enormous amounts of data using modern techniques of multivariate analysis [6]. Since 2001, a large number of scientific articles have appeared, relating to food authentication using new or existing analytical techniques in combination with multivariate data analysis. However, it has to be noted that the adulteration practices are also very contemporary and in constant development, with constant interest in surpassing the power of the established analytical methods of their discovery [14].

This chapter represents a thorough overview of the analytical methods employing a GC technique that are dealing with authentication and adulteration detection of various types of foodstuffs. The methods described are published in scientific literature in the last two decades.

2. Authentication and adulteration detection in various food products

2.1 Olive oil and other edible vegetable oils

Edible vegetable oils represent a matrix which is usually analyzed with the application of GC. That is why there are a large number of papers dealing with authentication and adulteration detection in this type of food, using GC. Among them, extra virgin and virgin olive oils are definitely the most investigated. The suggested analytical methods are focused on the determination of constituents in oil mixtures of high prices and quality, the discrimination of extra virgin olive oils from defected oils, the possibilities of the authentication of various edible oils and fats and the determination of geographical origin. Triacylglycerol composition, fatty acid composition, $^{13}C/^{12}C$ and $^{2}H/^{1}H$ ratios and enantiomeric distributions of certain compounds, and just in some cases volatile organics and phenolic compounds, are usually considered as discrimination factors. Considering that this kind of analysis provides a large amount of data, the recently published papers are almost exclusively coupling GC with various unsupervised and supervised techniques of multivariate chemometric data analysis. Among unsupervised principal component analysis is definitely the mostly used, and among supervised techniques and machine learning algorithms, there are many different described: LDA and SLDA, PLS-DA, OPLS-DA, SIMCA, ANN-MLP, R-SVM and OC-SVM and some other. **Table 1** lists chronological literature data on authentication and adulteration detection procedures of the most commonly investigated olive oil, and also edible oils of other plant species, and some examples of animal fats.

2.2 Honey and other bee products

The authenticity of honey and other bee products has two aspects. Authenticity in respect of production, i.e., to prevent adulteration by the addition of other food ingredients (various types of sugar syrups), and authenticity of botanical and geographical origin. The GC method for determining the addition of sugar syrups relies on carbohydrate profiling in combination with classical statistical procedures for data processing. However, methods for authentication of geographical and botanical origin of honey samples usually employ more complex sample preparations, such as solid-phase microextraction in a headspace mode, and more sophisticated instrumentation, such as multidimensional GC. These methods mostly rely on the analysis of volatile organic compounds and also usually involve the application of multivariate chemometric tools for data analysis—unsupervised and supervised pattern recognition techniques. Unsupervised techniques, PCA and HCA, are more commonly used, but some studies also report the application of supervised tools: LDA and SLDA, OPLS-DA, SIMCA and ANN-MLP. **Table 2** lists examples from literature data on authentication and adulteration detection procedures of honey and other bee products, such as beeswax, propolis and royal jelly.

2.3 Milk and dairy products

Authenticity of milk and dairy products, such as cheese and fermented milk, using GC, is usually based on the determination of fat content of samples: triacylglycerols and fatty acids. Therefore, it is usually enough to combine GC with FID, to perform a successful analysis. In some particular cases, MS or olfactometry is used (if the analytical method is based on determining volatile profiles of the samples). Methods described in the literature rarely use chemometric data analysis, in some cases PCA, LDA and PLS-DA, but rather rely on the application of classical statistics. Papers describing the authentication of milk and dairy products usually deal with discriminating organic from conventionally produced ones, discriminating samples according to geographical origin and according to the animal breed they are produced of. **Table 3** shows literature examples of authentication and adulteration detection practices in milk and dairy products, such as cheese.

Purpose of the study	Analytical technique	Chemometric technique	Ref.
Olive oil			
Discrimination of "Ligurian" from "non-Ligurian" olive oils	HS-SPME/GC-ITMS	LDA, ANN-MLP	[18]
Differentiation of monovarietal olive oils according to olive variety	HS-SPME/ GC × GC- TOF-MS HT-GC-ITMS HS-SPME/GC-MS GC-TOF-MS	PCA PCA, HCA, PLS- DA — PLS-DA	[19] [20] [21] [22]
Differentiation of extra virgin and virgin olive oils according to geographical origin (various regions in Spain, Italy)	HS-SPME/GC-MS HS-SPME/GC-MS GC-C/P-IRMS	SLDA — —	[23] [21] [24]
Discrimination of extra virgin olive oils from defected oils	HS-SPME/ GC × GC-MS	PCA, PLS-DA	[25]
Detection of extra virgin olive oil, virgin olive oil and olive oil adulteration (with various types of edible oils)	LC-GC-ITMS LC-chiral-GC- ITMS GC-MS SPME/ GC × GC-MS GC-MS SPME/GC-MS GC-FID GC-MS GC-MS	— — PCA — SIMCA, kNN, PLSR — PCA, PLS PCA, TFA, SIMCA, PLS OC- SVM	[26] [27] [28] [29] [30] [31] [32] [33] [34]
Other edible oils			
Discrimination of various vegetable oils according to botanical origin (sunflower, corn, sesame, soybean, olive, rapeseed, camellia, peanut, canola, palm, rice bran, coconut, grapeseed, hazelnut, walnut, apricot seed, red pepper seed, prikachberry, pumpkin)	GC-C-IRMS GC-C-IRMS HT-GC-FID GC-FID GC-FID HT-GC-MS GC-MS GC-MS GC-MS GC-MS	CDA — PCA PCA, KNN, CNN PCA, PLS SIMCA, PLS, GA-PLS PCA, PLS-DA, OPLS-DA PCA, HCA, RF LDA, GA-SVM	[35] [36] [37] [38] [32] [39] [40] [41] [42] [43]
Differentiation of almond oils according to almond variety	HS-SPME/GC-MS	SLDA	[44]
Detection of corn oil adulteration	GC-C-IRMS	—	[45]
Detection of flaxseed oil adulteration	GC-MS	PCA, R-SVM	[46]
Detection of sesame oil adulteration	GC-FID GC-FID GC-MS	— SVM OC-SVM	[47] [48] [49]
Authenticity and geographical origin of pumpkin seed oil	GC-FID GC-C-IRMS	PCA, RDA	[50]
Fats			
Authenticity of cocoa butter	LC-GC-MS	—	[51]
Discrimination of various edible oils and fats (pig, mutton, beef and chicken)	GC-MS	PCA, PLS-DA, OPLS-DA	[40]

Table 1.
Literature examples of authentication and adulteration detection procedures of olive oil and other edible oils and fats.

Purpose of the study	Analytical technique	Chemometric technique	Ref.
Honey			
Detection of the addition of sugar syrups to honey (high-fructose corn syrup and inverted syrup)	GC-FID GC-FID/MS GC-FID	PCA — —	[52] [53] [54]
Differentiation of four types of multifloral Portuguese honeys (produced in Madeira Island)	HS-SPME/GC-MS	PCA, SLDA	[55]
Authenticity of "Corsica" honey	HS-SPME/ GC × GC-TOF-MS HS-SPME/ GC × GC-TOF-MS	PCA, ANN-MPL LDA, SIMCA, SVM, DPLS	[56] [57]
Detection of honey adulteration with high-fructose inulin syrups	GC-MS	—	[58]
Authenticity of thistle honey	HD-SPME/GC-MS	—	[59]
Authenticity of botanical origin of unifloral chestnut (*Castanea sativa* L.) and eucalyptus (*Eucalyptus globulus* Labill.) honeys	GC-MS	—	[60]
Differentiation between lemon blossom honey (*Citrus limon*) and orange blossom honey (*Citrus* spp.)	GC-MS	PCA	[61]
Geographical origin identification of honey (samples from various regions of Greece; samples from various Mediterranean countries: Egypt, Greece, Morocco, Spain)	HS-GC-MS HS-SPME/GC-MS HS-GC-MS HS-SPME/GC-Q-TOF-MS HS-SPME/GC-MS SPME-GC/MS	— HCA, SLDA, kNN OPLS-DA, SIMCA, OPLS-HCA PCA LDA LDA	[62] [63] [64] [65] [66] [67]
Differentiation of honeys according to botanical origin: heather, raspberry, rape, alder buckthorn, lime, rosemary, chestnut, sunflower, acacia, thyme, orange, linden, amaranth, honeydew, citrus, *Gossypium*, rhododendron, alfalfa, white clover, carob, calden	HS-GC-MS SPME/GC-MS HS-GC-MS SPME/chiral-GC × GC-MS SPME/GC-MS/O SPME-GC/MS	— LDA OPLS-DA, SIMCA, OPLS-HCA — AHC, CA PCA, HCA	[62] [68] [64] [69] [70] [71]
Establishment of orange honey authenticity	SPME/GC-MS	—	[72]
Other bee products			
Authenticity of royal jelly; detection of the addition of sugar syrups	HR-GC	—	[73]
Characterization of traditional plant syrups from Spain, namely, palm honey (miel de palma), must syrup (arrope) and sugarcane honey (miel de caña)	GC-MS	—	[74]
Detection of adulterated beeswax from *Apis mellifera* L.	HT-GC-FID/MS HT-GC-FID/MS	— HCA, PCA, LDA	[75] [76]
Geographical origin identification of propolis	HS/GC-MS/O	PCA	[77]
Establishment of sugarcane honey authenticity	HS-SPME/GC-MS	PCA, LDA	[78]

Table 2.
Literature examples of authentication and adulteration detection procedures of honey and other bee products.

Purpose of the study	Analytical technique	Chemometric technique	Ref.
Milk			
Authenticity of goat milk	GC-FID	—	[79]
Differentiation between cow milk produced in the lowlands, mountains and highlands of Switzerland	HR-GC-FID	—	[80]
Authenticity of milk fat: detection of foreign fat in milk fat (such as pork lard, bovine tallow, fish oil, peanut oil, corn oil, olive oil, soy oil, sunflower oil, coconut fat)	GC-FID UFM-GC-FID GC-FID GC-FID GC-FID	— — LDA — —	[81] [82] [83] [84] [85]
Differentiation of milk produced under conventional and organic management	GC-FID GC-FID GC-MS	— — —	[86] [87] [88]
Differentiation of cow, goat, sheep, water buffalo, donkey, horse and camel milk	GC-FID	PCA	[89]
Determining the origin of milk samples: hay milk vs. conventional (silage) milk	GC-FID	PCA, PLS-DA	[90]
Dairy products			
Geographic origin of Emmental cheese	GC-FID HS-GC-FID/MS	— PCA	[91] [92]
Differentiation of Grana Padano, Parmigiano-Reggiano and Grana Trentino cheeses	GC-O	PCA	[93]
Examining foreign fat origin in cheese from cow milk fat	GC-FID	—	[94]
Differentiation between certified organic and conventional probiotic fermented milks	GC-FID	—	[95]
Quality control for Parmigiano-Reggiano cheese	GC-MS	PCA	[96]

Table 3.
Literature examples of authentication and adulteration detection procedures of milk and various dairy products.

2.4 Fruits and fruit-made beverages

Most of the papers dealing with fruit authenticity testing using GC are focused on determining discriminating factors that will enable discrimination of varieties of certain fruit species. These factors are mostly constituted of free and bound volatile compounds belonging to different chemical groups, namely, linear and branched esters, terpenes, alcohols and others. The paper published by Kurz et al. [97] is an exception, which is dealing with the analysis of neutral sugars of cell wall polysaccharide profiles of apricots, peaches and pumpkins using GC-FID. In some cases GC was also combined with other analytical techniques, thus enabling the wider spectra of chemical species to be included in the analysis, such as LC, in order to include nonvolatile carbohydrates, fatty acids and organic acids. The obtained data were mainly processed using multivariate data analysis techniques, such as HCA, PCA, PLS-DA, LDA and OPLS-DA. Older investigations usually do not include multivariate data analysis. Schmarr and Bernhardt [98] used image processing techniques in order to process the data obtained after comprehensive two-dimensional GC

Purpose of the study	Analytical technique	Chemometric technique	Ref.
Fruits			
Differentiation of blackcurrant (*Ribes nigrum* L.) berries	SPME/GC-FID	—	[97]
Differentiation of apricots (*Prunus armeniaca* L.), peaches (*Prunus persica* L.) and pumpkins (*Cucurbita* sp.)	GC-FID	—	[99]
Differentiation between *Passiflora* fruit species	HS-SPME/GC-MS	PCA	[100]
Discrimination of red grape varieties of southern Italy (Aglianico, Uva di Troia, Negroamaro, Primitivo)	GC-MS	PCA	[101]
Differentiation of apples, pears and quince fruit	HS-SPME/ GC × GC	—	[98]
Classification of apple varieties (Golden Delicious, Granny Smith, Pinova and Stark Delicious)	HS-SPME/GC-TOF-MS	PCA, PLS-DA	[102]
Differentiation between grape varieties: *Vitis vinifera*, *Vitis cinerea* and interspecific crosses	GC-MS	HCA	[103]
Differentiation of Chinese bayberry cultivars (*Myrica rubra*)	HS-SPME/GC-MS HS-SPME/GC-MS/O	PCA —	[104] [105]
Discrimination of nine passion fruits: yellow, purple, lemon, orange, pineapple, peach, melon, banana and tomato	HS-SPME/GC-MS	PCA, PLS-DA	[106]
Differentiation between Tanzanian grown fruits: mango, pineapple, jackfruit, baobab and tamarind	GC-MS	HCA, PCA	[107]
Discrimination of *Eugenia uniflora* L. biotypes	HS-SPME/GC-MS	PCA, HCA	[108]
Differentiation between date palm fruit (*Phoenix dactylifera* L.) varieties from Egypt	SPME/GC-MS	PCA, HCA, OPLS-DA	[109]
Characterization of organic oranges (*Citrus sinensis* L. Osbeck)	HS-SPM/GC-MS	PLS-DA	[110]
Differentiation of sun-dried raisins made from different grape varieties	HS-SPME/GC-TOF-MS	—	[111]
Differentiation between citrus species: mandarin, sweet orange, sour orange, papeda, pummelo, lemon, *Fortunella* Swingle	GC-MS	HCA	[112]
Differentiation of apple cultivars from geographical origin and growing conditions (organic and conventional)	HS-SPME/GC-MS	PLS-DA	[113]
Differentiation of Chinese Jujube varieties	HS-SPME/GC-MS	HCA	[114]
Fruit beverages			
Detecting adulteration of blackcurrant juice	GC-FID	—	[115]
Authentication of apple and orange juice	GC-FID	—	[116]
Detecting the addition of aromas to fruit beverages	SPME/chiral-GC-MS	—	[117]
Citrus juice classification (lemon, grapefruit, mandarin, orange, lime)	HS-SPME/GC-MS	LDA	[118]
Assessment of premium organic orange juices authenticity	HS-SPME/GC-MS	PLS-DA	[119]

Table 4.
Literature examples of authentication and adulteration detection procedures of various fruits and fruit juices.

analysis. **Table 4** chronologically lists some literature examples on authentication and adulteration procedures of various fruit species and fruit-made juices.

2.5 Cereals and bakery products

Cereals, pseudocereals, flours and bread, as mostly used bakery products in human nutrition, are usually differentiated according to varietal, botanical or geographical origin by combining GC analysis with chemometric processing of the obtained data. The chemical compounds that have the role of discriminating factors usually involve small molecules, such as simple soluble sugars and free fatty acids. Chemometric methods involve most often exploratory data analysis techniques, such as PCA, PCO and HCA, but in some cases also classification methods of LDA and QDA were applied to measure the classification and prediction abilities. **Table 5** chronologically lists some literature examples of authentication and adulteration detection practices of cereals, flour and the most commonly used bakery product in human nutrition-bread.

2.6 Meat, fish and seafood

The studies of authenticity of seafood and meat products using a GC technique usually focus on the determination of freshness of a seafood or meat product. Chemometric techniques, such as PCA, were able to successfully discriminate between fresh samples, deteriorated samples and gradually decaying samples of seafood, and ANN were employed in order to classify samples of fresh meat, frozen-thawed meat and spoiled meat. The PCA of gas chemometric fingerprints was able to show separation not only between oyster species but also between oysters originating from different cultivation areas, as well as oysters harvested at

Purpose of the study	Analytical technique	Chemometric technique	Ref.
Cereals			
Differentiation between *Triticum durum* and *Triticum aestivum*	GC-FID	PCA, LDA, QDA	[120]
Differentiation between hexaploid (*T. aestivum*, *T. spelta*) and tetraploid (*T. durum*, *T. dicoccon*) wheats	GC-MS	—	[121]
Classifications of cereals (wheat and corn) used in DDGS material by geographical and botanical origin	GC-FID	PLS-DA	[122]
Flour			
Differentiation of corn and small grain flour (wheat, rye, triticale, barley, oats)	GC-MS	HCA, PCO HCA, PCA	[123] [124]
Differentiation of corn and oat flour, from other small grains (wheat, barley, triticale, rye)	GC-MS	HCA, PCO	[125]
Differentiation of flours of corn, spelt, buckwheat, amaranth and small grains (wheat, rye, triticale, oats, barley)	GC-MS	HCA, PCA	[126]
Bakery products			
The content of buckwheat flour in wheat bread	GC-MS	HCA HCA, PCA	[127] [128]

Table 5.
Literature examples of authentication and adulteration detection procedures of cereals, flour and bakery products.

different time intervals. There was only one paper found in the literature that deals with differentiation of meat according to the breed origin. The PCA was successfully applied to discriminate between samples of pork, chicken, beef and mutton meat. **Table 6** represents a chronological list of examples of authentication and adulteration detection procedures of various types of meat, fish and seafood.

Purpose of the study	Analytical technique	Chemometric technique	Ref.
Fish and seafood			
Differentiation between fresh and deteriorated oyster *Crassostrea gigas*	HS-SPME/GC-MS	PCA	[129]
Differentiation between fresh and frozen-thawed cultured gilthead sea bream fish (*Sparus aurata*)	SPME/GC-MS	—	[130]
Razor clam (*Sinonovacula constricta* Lamarck), redspot swimming crab (*Portunus sanguinolentus* Herbst) and prawn (*Penaeus japonicus* (Bate; Kuruma prawn)) freshness determination	HS-SPME/GC-MS	PCA	[131]
Differentiation of European flat oyster (*Ostrea edulis*) and Pacific cupped oyster (*Crassostrea gigas*): species, different cultivation areas, different time intervals of harvest	GC-FID GC-MS	PCA	[132]
Meat			
Halal authentication of pork meat	HS/GC-MS	PCA	[133]
Differentiation of fresh and frozen pork	UFGC	PCA, ANN	[134]

Table 6.
Literature examples of authentication and adulteration detection procedures of meat products and seafood.

Purpose of the study	Analytical technique	Chemometric technique	Ref.
Coffee			
Differentiation between arabica (*Coffea arabica* Linn.) and robusta (*Coffea canephora* Pierre ex Froehner var. *robusta*) coffees, either in green or in roasted stage	HR-GC-FID	HCA, CVA, DA	[135]
Determining the geographical origin of coffee samples	HS-SPME/GC-TOF-MS	PCA	[136]
Tea			
Differentiation of *Echinacea* species (*E. angustifolia*, *E. pallida*, *E. purpurea*)	GC-MS	HCA, PCA, LDA	[137]
Discrimination of oolong tea (*Camellia sinensis*) varieties	HS-SPME/GC-MS	PCA, HCA, SLDA	[138]
Discrimination of two roselle (*Hibiscus sabdariffa*) flower cultivars	SPME/GC-MS	PCA, HCA, OPLS-DA	[139]
Discrimination of different teas (*Camellia sinensis*)	HS-SPME/chiral-GC-MS	HCA, PLS-DA	[140]
Discrimination of American ginseng (*Panax quinquefolius* L.) and Asian ginseng (*Panax ginseng* Meyer)	GC-MS	PCA, PLS	[141]

Table 7.
Literature examples of authentication and adulteration detection procedures of coffee and tea.

2.7 Coffee and tea

Differentiation of coffee samples is based mostly on fatty acid profiles and volatile and semi-volatile compounds (organic acids, sugars, terpenoids). Differentiations of various tea plants were based exclusively on volatile components. In order to enable differentiations and classifications of investigated samples of beverages, the data obtained after GC analysis were combined with various chemometric techniques: HCA, PCA, SLDA and OPLS-DA. **Table** 7 represents chronological literature data on the authentication and adulteration detection procedures of coffee and tea from various plant species.

3. Conclusions

Gas chromatograph, as a common instrument in most analytical laboratories worldwide, can be successfully applied in authentication and fraud detection procedures of various food and beverage products, such as olive oil and other edible vegetable oils, honey and other bee products, milk and dairy products, cereals and bakery products, meat, fish and seafood, as well as coffee and tea. In this manner, gas chromatograph is coupled to flame ionization detector or single/tandem mass spectrometers. It can be concluded that utilization of a GC device in further development of authentication methodologies could provide us with meaningful results, thus representing a significant contribution to this emerging field in the future.

Acknowledgements

The authors would like to acknowledge the support from the Ministry of Education, Science and Technological Development of the Republic of Serbia.

Conflict of interest

The authors declare that they have no conflict of interest.

Acronyms and abbreviations

AHC	agglomerative hierarchical clustering
ANN	artificial neural networks
C-IRMS	combustion isotope ratio mass spectrometry
CA	correspondence analysis
CDA	canonical discriminant analysis
CNN	counterpropagation neural network
CVA	canonical variates analysis
DA	discriminant analysis
DPLS	discriminant partial least squares
FID	flame ionization detector
GA	genetic algorithm
GC	gas chromatography
GC-chiral GC	fast multiple heart-cut enantioselective multidimensional gas chromatography

chiral GC × GC	enantioselective comprehensive two-dimensional gas chromatography
HCA	hierarchical cluster analysis
HR	high resolution
HS	headspace
HT	high temperature
IR	isotope ratio
IT	ion-trap
kNN	k-nearest neighbors
KNN	Kohonen neural network
LC	liquid chromatography
LDA	linear discriminant analysis
MLP	multilayer perceptron
MCOCPLS	Monte Carlo one-class partial least squares
MS	mass spectrometry
O	olfactometry
OC-SVM	one-class support vector machine
OPLS-DA	orthogonal projections to latent structures discriminant analysis
P-IRMS	pyrolysis isotope ratio mass spectrometry
PCA	principal component analysis
PCoA	principal coordinate analysis
PLS	principal least squares regression
Q-TOF-MS	quadrupole accurate mass time-of-flight mass spectrometry
QDA	quadratic discriminant analysis
R-SVM	recursive support vector machine
RDA	regularized discriminant analysis
RF	random forests
SIMCA	soft independent modeling of class analogy
SLDA	stepwise linear discriminant analysis
SPME	solid-phase microextraction
SVM	support vector machine
TOF	time-of-flight
UF	ultrafast module

Author details

Kristian Pastor*, Marijana Ačanski and Djura Vujić
Faculty of Technology Novi Sad, University of Novi Sad, Novi Sad, Serbia

*Address all correspondence to: pastor@tf.uns.ac.rs

References

[1] Oliveri P, Downey G. Multivariate class modeling for the verification of food-authenticity claims. Trends in Analytical Chemistry. 2012;**35**:74-86

[2] Danezis GP, Tsagkaris AS, Camin F, Brusic V, Georgiou CA. Food authentication: Techniques, trends & emerging approaches. Trends in Analytical Chemistry. 2016a;**85**:123-132

[3] Accum F. A Treatise on Adulterations of Food and Culinary Poisons. London, UK: Longman; 1820

[4] Manning L, Soon JM. Developing systems to control food adulteration. Food Policy. 2014;**49**:23-32

[5] Spink J, Moyer DC. Defining the public health threat of food fraud. Journal of Food Science. 2011;**76**:157-163

[6] Borras E, Ferre J, Boque R, Mestres M, Acena L, Busto O. Data fusion methodologies for food and beverage authentication and quality assessment—A review. Analytica Chimica Acta. 2015;**891**:1-14

[7] Cuadros-Rodríguez L, Ruiz-Samblas C, Valverde-Som L, Perez-Castano E, Gonzalez-Casado A. Chromatographic fingerprinting: An innovative approach for food 'identitation' and food authentication—A tutorial. Analytica Chimica Acta. 2016;**909**:9-23

[8] Cubero-Leon E, Peñalver R, Maquet A. Review on metabolomics for food authentication. Foodservice Research International. 2014;**60**:95-107

[9] Danezis GP, Tsagkaris AS, Brusic V, Georgiou CA. Food authentication: State of the art and prospects. Current Opinion in Food Science. 2016b;**10**:22-31

[10] WDS. Metabolomics: Applications to food science and nutrition research. Trends in Food Science and Technology. 2008;**19**:482-493

[11] Moore JC, Spink J, Development LM. Application of a database of food ingredient fraud and economically motivated adulteration from 1980 to 2010. Journal of Food Science. 2012;**77**: 118-126

[12] Manning L. Food fraud: Policy and food chain. Current Opinion in Food Science. 2016;**10**:16-21

[13] European Commission Regulation (EC) No 178/2002 of the European Parliament and of the council of 28 January 2002 laying down the general principles and requirements of food law, establishing the European Food Safety Authority and laying down procedures in matters of food safety. Official Journal of the European Communities. 2002;**L31**:1-24

[14] Reid LM, O'Donnellb CP, Downey G. Recent technological advances for the determination of food authenticity. Trends in Food Science and Technology. 2006;**17**:344-353

[15] European Commission Regulation (EU) No 1169/2011 of the European Parliament and of the council of 25 October 2011 on the provision of food information to consumers. Official Journal of the European Communities. 2011;**L304**(/18):18-62

[16] Corrado G. Advances in DNA typing in the agro-food supply chain. Trends in Food Science and Technology. 2016;**52**:80-89

[17] Gallo M, Ferranti P. The evolution of analytical chemistry methods in foodomics. Journal of Chromatography. A. 2016;**1428**:3-15

[18] Cajka T, Riddellova K, Klimankova E, Cerna M, Pudil F,

Hajslova J. Traceability of olive oil based on volatiles pattern and multivariate analysis. Food Chemistry. 2010;**121**: 282-289

[19] Torres Vaz-Freire L, Gomes da Silva MDR, Costa Freitas AM. Comprehensive two-dimensional gas chromatography for fingerprint pattern recognition in olive oils produced by two different techniques in Portuguese olive varieties *Galega Vulgar*, *Cobranc¸osa* e *Carrasquenha*. Analytica Chimica Acta. 2009;**633**:263-270

[20] Ruiz-Samblás C, Cuadros-Rodríguez L, González-Casado A, de Paula Rodríguez García F, de la Mata-Espinosa P, Bosque-Sendra JM. Multivariate analysis of HT/GC-(IT)MS chromatographic profiles of triacylglycerol for classification of olive oil varieties. Analytical and Bioanalytical Chemistry. 2011;**399**:2093-2103

[21] Cecchi T, Alfei B. Volatile profiles of Italian monovarietal extra virgin olive oils via HS-SPME-GC-MS: Newly identified compounds, flavors molecular markers, and terpenic profile. Food Chemistry. 2013;**141**:2025-2035

[22] Bajoub A, Pacchiarotta T, Hurtado-Fernández E, Olmo-García L, García-Villalba R, Fernández-Gutiérrez A, et al. Comparing two metabolic profiling approaches (liquid chromatography and gas chromatography coupled to mass spectrometry) for extra-virgin olive oil phenolic compounds analysis: A botanical classification perspective. Journal of Chromatography. A. 2016; **1428**:267-279

[23] Pizarro C, Rodriguez-Tecedor S, Perez-del-Notario N, Gonzalez-Saiz JM. Recognition of volatile compounds as markers in geographical discrimination of Spanish extra virgin olive oils by chemometric analysis of non-specific chromatography volatile profiles. Journal of Chromatography. A. 2011; **1218**:518-523

[24] Paolini M, Bontempo L, Camina F. Compound-specific δ13C and δ2H analysis of olive oil fatty acids. Talanta. 2017;**174**:38-43

[25] Purcaro G, Cordero C, Liberto E, Bicchi C, Contea LS. Toward a definition of blueprint of virgin olive oil by comprehensive two-dimensional gas chromatography. Journal of Chromatography. A. 2014;**1334**:101-111

[26] Blanch GP, del Mar Caja M, del Castillo MLR, Herraiz M. Comparison of different methods for the evaluation of the authenticity of olive oil and hazelnut oil. Journal of Agricultural and Food Chemistry. 1998;**46**:3153-3157

[27] del Castillo MLR, Herraiz M, Blanch GP. Determination of the enantiomeric composition of γ-lactones in edible oils by on-line coupled high performance liquid chromatography and gas chromatography. Journal of Agricultural and Food Chemistry. 2000; **48**:1186-1190

[28] Dourtoglou VG, Dourtoglou T, Antonopoulos A, Stefanou E, Lalas S, Poulos C. Detection of olive oil adulteration using principal component analysis applied on total and Regio FA content. JAOCS. 2003;**80**:203-208

[29] Flores G, del Castillo MLR, Blanch GP, Herraiz M. Detection of the adulteration of olive oils by solid phase microextraction and multidimensional gas chromatography. Food Chemistry. 2006;**97**:336-342

[30] Priego Capote F, Ruiz Jiménez J, Luque de Castro MD. Sequential (step-by-step) detection, identification and quantitation of extra virgin olive oil adulteration by chemometric treatment of chromatographic profiles. Analytical and Bioanalytical Chemistry. 2007;**388**: 1859-1865

[31] Sakouhi F, Absalon C, Flamini G, Cioni PL, Kallel H, Boukhchina S. Lipid

components of olive oil from Tunisian cv. Sayali: Characterization and authenticity. Comptes Rendus Biologies. 2010;**333**:642-648

[32] Rohman A, Che Man YB. Authentication of extra virgin olive oil from sesame oil using FTIR spectroscopy and gas chromatography. International Journal of Food Properties. 2011;**15**:1309-1318

[33] Monfreda M, Gobbi L, Grippa A. Blends of olive oil and seeds oils: Characterisation and olive oil quantification using fatty acids composition and chemometric tools. Food Chemistry. 2014;**145**:584-592

[34] Zhang L, Yuan Z, Li P, Wang X, Mao J, Zhang Q, et al. Targeted multivariate adulteration detection based on fatty acid profiles and Monte Carlo one-class partial least squares. Chemometrics and Intelligent Laboratory Systems. 2017;**169**:94-99

[35] Kelly S, Parker I, Sharman M, Dennis J, Goodall I. Assessing the authenticity of single oils using fatty acid stable carbon (13C/12C). Food Chemistry. 1997;**59**:181-186

[36] Woodbury SE, Evershed RP, Rossell JB. δ13C analyses of vegetable oil fatty acid components, determined by gas chromatography-combustion-isotope ratio mass spectrometry, after saponification or regiospecific hydrolysis. Journal of Chromatography. A. 1998;**805**:249-257

[37] Park JR, Lee DS. Detection of adulteration in olive oils using triacylglycerols compositions by high temperature gas chromatography. Bulletin of the Korean Chemical Society. 2003;**24**:527-530

[38] Brodnjak-Voncina D, Cencic Kodbab Z, Novic M. Multivariate data analysis in classification of vegetable oils characterized by the content of fatty acids. Chemometrics and Intelligent Laboratory Systems. 2005;**75**:31-43

[39] Ruiz-Samblás C, Marini F, Cuadros-Rodríguez L, González-Casado A. Quantification of blending of olive oils and edible vegetable oils by triacylglycerol fingerprint gas chromatography and chemometric tools. Journal of Chromatography B. 2012;**910**:71-77

[40] Fang G, Goh JY, Tay M, Lau HF, Li SFY. Characterization of oils and fats by 1H NMR and GC/MS fingerprinting: Classification, prediction and detection of adulteration. Food Chemistry. 2013; **138**:1461-1469

[41] Zhang L, Li P, Sun X, Wang X, Xu B, Wang X, et al. Adulteration detection of vegetable oils based on fatty acid profiles. Journal of Agricultural and Food Chemistry. 2014;**62**:8745-8751

[42] Troya F, Lerma-García MJ, Herrero-Martínez JM, Simó-Alfonso EF. Classification of vegetable oils according to their botanical origin using n-alkane profiles established by GC-MS. Food Chemistry. 2015;**167**:36-39

[43] Li X, Kong W, Shi W, Shen Q. A combination of chemometrics methods and GC-MS for the classification of edible vegetable oils. Chemometrics and Intelligent Laboratory Systems. 2016; **155**:145-150

[44] Beltrán A, Ramos M, Grané N, Martín ML, Garrigós MC. Monitoring the oxidation of almond oils by HS SPME-GC-MS and ATR-FTIR: Application of volatile compounds determination to cultivar authenticity. Food Chemistry. 2011;**126**:603-609

[45] Woodbury SE, Evershed RP. Detection of vegetable oil adulteration using gas chromatography combustion/ lsotope ratio mass spectrometry. Analytical Chemistry. 1995;**67**: 2685-2690

[46] Sun X, Zhang L, Li P, Xu B, Ma F, Zhang Q, et al. Fatty acid profiles based adulteration detection for flaxseed oil by gas chromatography mass spectrometry. LWT—Food Science and Technology. 2015;**63**:430-436

[47] Park YW, Chang P-S, Lee JH. Application of triacylglycerol and fatty acid analyses to discriminate blended sesame oil with soybean oil. Food Chemistry. 2010;**123**:377-383

[48] Peng D, Bi Y, Ren X, Yang G, Sun S, Wang X. Detection and quantification of adulteration of sesame oils with vegetable oils using gas chromatography and multivariate data analysis. Food Chemistry. 2015;**188**:415-421

[49] Zhang L, Huang X, Li P, Naf W, Jiang J, Mao J, et al. Multivariate adulteration detection for sesame oil. Chemometrics and Intelligent Laboratory Systems. 2017;**161**:147-150

[50] Potocnik T, Ogrinc N, Potocnik D, Košir IJ. Fatty acid composition and δ13C isotopic ratio characterisation of pumpkin seed oil. Journal of Food Composition and Analysis. 2016;**53**: 85-90

[51] Kamm W, Dionisi F, Fay L-B, Hischenhuber C, Schmarr H-G, Engel K-H. Analysis of steryl esters in cocoa butter by on-line liquid chromatography-gas chromatography. Journal of Chromatography. A. 2001; **918**:341-349

[52] Cotte JF, Casabianca H, Chardon S, Lheritier J, Grenier-Loustalot MF. Application of carbohydrate analysis to verify honey authenticity. Journal of Chromatography. A. 2003;**1021**:145-155

[53] Ruiz-Matute AI, Soria AC, Martınez-Castro I, Sanz ML. New methodology based on GC-MS to detect honey adulteration with commercial syrups. Journal of Agricultural and Food Chemistry. 2007;**55**:7264-7269

[54] Kaškonienė V, Venskutonis P, Čeksterytė V. Sugar analysis for authenticity evaluation of honey in Lithuanian market. Acta Alimentaria. 2011;**40**:205

[55] Pontes M, Marques JC, Camara JS. Screening of volatile composition from Portuguese multifloral honeys using headspace solid-phase microextraction-gas chromatography-quadrupole mass spectrometry. Talanta. 2007;**74**:91-103

[56] Cajka T, Hajslova J, Pudil F, Riddellova K. Traceability of honey origin based on volatiles pattern processing by artificial neural networks. Journal of Chromatography. A. 2009; **1216**:1458-1462

[57] Stanimirova I, Ustun B, Cajka T, Riddelova K, Hajslova J, Buydens LMC, et al. Tracing the geographical origin of honeys based on volatile compounds profiles assessment using pattern recognition techniques. Food Chemistry. 2010;**118**:171-176

[58] Ruiz-Matute AI, Rodrıguez-Sanchez S, Sanz ML, Martınez-Castro I. Detection of adulterations of honey with high fructose syrups from inulin by GC analysis. Journal of Food Composition and Analysis. 2010a;**23**:273-276

[59] Bianchi F, Mangia A, Mattarozzi M, Musci M. Characterization of the volatile profile of thistle honey using headspace solid-phase microextraction and gas chromatography-mass spectrometry. Food Chemistry. 2011; **129**:1030-1036

[60] Alissandrakis E, Tarantilis PA, Pappas C, Harizanis PC, Polissiou M. Investigation of organic extractives from unifloral chestnut (*Castanea sativa* L.) and eucalyptus (*Eucalyptus globulus* Labill.) honeys and flowers to identification of botanical marker compounds. LWT- Food Science and Technology. 2011;**44**:1042-1051

[61] Escriche I, Kadar M, Juan-Borrás M, Domenech E. Using flavonoids, phenolic compounds and headspace volatile profile for botanical authentication of lemon and orange honeys. Foodservice Research International. 2011;**44**:1504-1513

[62] Radovic BS, Careri M, Mangia A, Musci M, Gerboles M, Anklam E. Contribution of dynamic headspace GC-MS analysis of aroma compounds to authenticity testing of honey. Food Chemistry. 2001;**72**:511-520

[63] Baroni MVN, Nores ML, Del Pilar Diäaz M, Chiabrando GA, Fassano JB, Costa C, et al. Determination of volatile organic compound patterns characteristic of five unifloral honey by solid-phase microextraction-gas chromatography-mass spectrometry coupled to chemometrics. Journal of Agricultural and Food Chemistry. 2006; **54**:7235-7724

[64] Aliferis KA, Tarantilis PA, Harizanis PC, Alissandrakis E. Botanical discrimination and classification of honey samples applying gas chromatography/mass spectrometry fingerprinting of headspace volatile compounds. Food Chemistry. 2010;**121**: 856-862

[65] Moniruzzaman M, Rodríguez I, Ramil M, Cela R, Sulaiman SA, Gan SH. Assessment of gas chromatography time-of-flight accurate mass spectrometry for identification of volatile and semi-volatile compounds in honey. Talanta. 2014;**129**:505-515

[66] Karabagias IK, Badeka A, Kontakos S, Karabournioti S, Kontominas MG. Characterisation and classification of Greek pine honeys according to their geographical origin based on volatiles, physicochemical parameters and chemometrics. Food Chemistry. 2014;**146**:548-557

[67] Karabagias IK, Louppis AP, Karabournioti S, Kontakos S, Papastephanou C, Kontominas MG. Characterization and geographical discrimination of commercial *Citrus* spp. honeys produced in different Mediterranean countries based on minerals, volatile compounds and physicochemical parameters, using chemometrics. Food Chemistry. 2017;**217**:445-455

[68] Senyuva HZ, Gilbert J, Silici S, Charlton A, Dal C, Gurel N, et al. Profiling Turkish honeys to determine authenticity using physical and chemical characteristics. Journal of Agricultural and Food Chemistry. 2009; **57**:3911-3919

[69] Španik I, Pazitna A, Šiška P, Szolcsanyi P. The determination of botanical origin of honeys based on enantiomer distribution of chiral volatile organic compounds. Food Chemistry. 2014;**158**:497-503

[70] Seisonen S, Kivima E, Vene K. Characterisation of the aroma profiles of different honeys and corresponding flowers using solid-phase microextraction and gas chromatography-mass spectrometry/ olfactometry. Food Chemistry. 2015; **169**:34-40

[71] Tahir HE, Xiaobo Z, Xiaowei H, Jiyong S, Mariod AA. Discrimination of honeys using colorimetric sensor arrays, sensory analysis and gas chromatography techniques. Food Chemistry. 2016;**206**:37-43

[72] Verzera A, Tripodi G, Condurso C, Dima G, Marra A. Chiral volatile compounds for the determination of orange honey authenticity. Food Control. 2014;**39**:237-243

[73] Lercker G, Savioli S, Vecchi MA, Sabatini AG, Nanetti A, Piana L. Carbohydrate determination of Royal Jelly by high resolution gas chromatography (HRGC). Food Chemistry. 1986;**19**:255-264

[74] Ruiz-Matute AI, Soria AC, Sanz ML, Martınez-Castro I. Characterization of traditional Spanish edible plant syrups based on carbohydrate GC-MS analysis. Journal of Food Composition and Analysis. 2010B;**23**:260-263

[75] Serra Bonvehi J, Orantes Bermejo FJ. Detection of adulterated commercial Spanish beeswax. Food Chemistry. 2012;**132**:642-648

[76] Maia M, Nunes FM. Authentication of beeswax (*Apis mellifera*) by high-temperature gas chromatography and chemometric analysis. Food Chemistry. 2013;**136**:961-968

[77] Cheng H, Qin ZH, Guo XF, Hu XS, Wu JH. Geographical origin identification of propolis using GC-MS and electronic nose combined with principal component analysis. Foodservice Research International. 2013;**51**:813-822

[78] Silva P, Freitas J, Silva CL, Perestrelo R, Nunes FM, Camara JS. Establishment of authenticity and typicality of sugarcane honey based on volatile profile and multivariate analysis. Food Control. 2017;**73**: 1176-1188

[79] Fontecha J, Díaz V, Fraga MJ, Juárez M. Triglyceride analysis by gas chromatography in assessment of authenticity of goat milk fat. Journal of the American Oil Chemists' Society. 1998;**75**:1893-1896

[80] Collomb M, Butikofer U, Sieber R, Jeangros B, Bosset J-O. Composition of fatty acids in cow's milk fat produced in the lowlands, mountains and highlands of Switzerland using high-resolution gas chromatography. International Dairy Journal. 2002;**12**:649-659

[81] Destaillats F, de Wispelaere M, Joffre F, Golay P-A, Hug B, Giuffrida F, et al. Authenticity of milk fat by fast analysis of triacylglycerols. Application to the detection of partially hydrogenated vegetable oils. Journal of Chromatography. A. 2006;**1131**:227-234

[82] Povolo M, Pelizzola V, Contarini G. Directly resistively heated-column gas chromatography for the evaluation of cow milk fat purity. European Journal of Lipid Science and Technology. 2008; **110**:1050-1057

[83] Gutiérrez R, Vega S, Díaz G, Sánchez J, Coronado M, Ramírez A, et al. Detection of non-milk fat in milk fat by gas chromatography and linear discriminant analysis. Journal of Dairy Science. 2009;**92**:1846-1855

[84] Molkentin J. The effect of cheese ripening on milkfat composition and the detection of fat from non-dairy origin. International Dairy Journal. 2013;**33**: 16-21

[85] Park J-M, Kim N-K, Yang C-Y, Moon K-W, Kim J-M. Determination of the authenticity of dairy products on the basis of fatty acids and triacylglycerols content using GC analysis. Korean Journal of Food Science and Analysis. 2014;**34**:316-324

[86] Butler G, Stergiadis S, Seal C, Eyre M, Leifert C. Fat composition of organic and conventional retail milk in Northeast England. Journal of Dairy Science. 2011;**94**:24-36

[87] Capuano E, Gravink R, Boerrigter-Eenling R, van Ruth SM. Fatty acid and triglycerides profiling of retail organic, conventional and pasture milk: Implications for health and authenticity. International Dairy Journal. 2015;**42**: 58-63

[88] Schwendel BH, Morel PCH, Wester TJ, Tavendale MH, Deadman C, Fong B, et al. Fatty acid profile differs between organic and conventionally produced cow milk independent of season or milking time. Journal of Dairy Science. 2015;**98**:1411-1425

[89] Smiddy MA, Huppertz T, van Ruth SM. Triacylglycerol and melting profiles of milk fat from several species. International Dairy Journal. 2012;**24**: 64-69

[90] Werteker M, Huber S, Kuchling S, Rossmann B, Schreiner M. Differentiation of milk by fatty acid spectra and principal component analysis. Measurement. 2017;**98**:311-320

[91] Pillonel L, Collomb M, Tabacchi M, Bosset JO. Analytical methods for the determination of the geographic origin of Emmental cheese. Free fatty acids, triglycerides and fatty acid composition of cheese fat. Mitteilungen aus Lebensmitteluntersuchung und Hygiene. 2002;**93**:217-231

[92] Pillonel L, Ampuero S, Tabacchi R, Bosset JO. Analytical methods for the determination of the geographic origin of Emmental cheese: Volatile compounds by GC/MS-FID and electronic nose. European Food Research and Technology. 2003;**216**: 179-183

[93] Boscaini E, van Ruth S, Biasioli F, Gasperi F, Mark TD. Gas chromatography-olfactometry (GC-O) and proton transfer reaction-mass spectrometry (PTR-MS) analysis of the flavor profile of Grana Padano, Parmigiano Reggiano, and Grana Trentino cheeses. Journal of Agricultural and Food Chemistry. 2003;**51**:1782-1790

[94] Fontecha J, Mayo I, Toledano G, Juarez M. Triacylglycerol composition of protected designation of origin cheeses during ripening. Authenticity of milk fat. Journal of Dairy Science. 2006; **89**:882-887

[95] Florence ACR, Béal C, Silva RC, Bogsan CSB, Pilleggi ALOS, Gioielli LA, et al. Fatty acid profile, trans-octadecenoic, α-linolenic and conjugated linoleic acid contents differing in certified organic and conventional probiotic fermented milks. Food Chemistry. 2012;**135**:2207-2214

[96] Marseglia A, Caligiani A, Comino L, Righi F, Quarantelli A, Palla G. Cyclopropyl and x-cyclohexyl fatty acids as quality markers of cow milk and cheese. Food Chemistry. 2013;**140**: 711-716

[97] Kurz C, Carle R, Schieber A. Characterisation of cell wall polysaccharide profiles of apricots (*Prunus armeniaca* L.), peaches (*Prunus persica* L.), and pumpkins (*Cucurbita* sp.) for the evaluation of fruit product authenticity. Food Chemistry. 2008; **106**:421-430

[98] Schmarr H-G, Bernhardt J. Profiling analysis of volatile compounds from fruits using comprehensive two-dimensional gas chromatography and image processing techniques. Journal of Chromatography. A. 2010;**1217**:565-574

[99] del Castillo MLR, Dobson G. Varietal differences in terpene composition of blackcurrant (*Ribes nigrum* L) berries by solid phase microextraction/gas chromatography. Journal of the Science of Food and Agriculture. 2002;**82**:1510-1515

[100] Pontes M, Marques JC, Câmara JS. Headspace solid-phase microextraction-gas chromatography quadrupole mass spectrometric methodology for the establishment of the volatile composition of *Passiflora* fruit species. Microchemical Journal. 2009;**93**:1-11

[101] Tamborra P, Esti M. Authenticity markers in *Aglianico*, *Uva di Troia*, Negroamaro and Primitivo grapes. Analytica Chimica Acta. 2010;**660**: 221-226

[102] Aprea E, Gika H, Carlin S, Theodoridis G, Vrhovsek U, Mattivi F. Metabolite profiling on apple volatile content based on solid phase microextraction and gas-chromatography

time of flight mass spectrometry. Journal of Chromatography. A. 2011; **1218**:4517-4524

[103] Ghaste M, Narduzzi L, Carlin S, Vrhovsek U, Shulaev V, Mattivi F. Chemical composition of volatile aroma metabolites and their glycosylated precursors that can uniquely differentiate individual grape cultivars. Food Chemistry. 2015;**188**:309-319

[104] Cheng H, Chen J, Li X, Pan J, Xue SY, Liu D, et al. Differentiation of the volatile profiles of Chinese bayberry cultivars during storage by HS-SPME-GC/MS combined with principal component analysis. Postharvest Biology and Technology. 2015a;**100**: 59-72

[105] Cheng H, Chen J, Chen S, Wu D, Liu D, Ye X. Characterization of aroma-active volatiles in three Chinese bayberry (*Myrica rubra*) cultivars using GC-MS-olfactometry and an electronic nose combined with principal component analysis. Foodservice Research International. 2015b;**72**:8-15

[106] Porto-Figueira P, Freitas A, Cruz CJ, Figueira J, Câmara JC. Profiling of passion fruit volatiles: An effective tool to discriminate between species and varieties. Foodservice Research International. 2015;**77**:408-418

[107] Khakimov B, Mongi RJ, Sørensen KM, Ndabikunze BK, Chove BE, Engelsen SB. A comprehensive and comparative GC-MS metabolomics study of nonvolatiles in Tanzanian grown mango, pineapple, jackfruit, baobab and tamarind fruits. Food Chemistry. 2016;**213**:691-699

[108] Mesquita PRR, Nunes EC, dos Santos FN, Bastos LP, MAPC C, de Rodrigues MF, et al. Discrimination of *Eugenia uniflora* L. biotypes based on volatile compounds in leaves using HS-SPME/GC-MS and chemometric analysis. Microchemical Journal. 2017; **130**:79-87

[109] Khalil MNA, Fekry MI, Farag MA. Metabolome based volatiles profiling in 13 date palm fruit varieties from Egypt via SPME GC-MS and chemometrics. Food Chemistry. 2017;**217**:171-181

[110] Cuevas FJ, Moreno-Rojas JM, Ruiz-Moreno MJ. Assessing a traceability technique in fresh oranges (*Citrus sinensis* L. Osbeck) with an HS-SPME-GC-MS method. Towards a volatile characterisation of organic oranges. Food Chemistry. 2017;**221**: 1930-1938

[111] Wang D, Duan CQ, Shi Y, Zhu BQ, Javed HU, Wang J. Free and glycosidically bound volatile compounds in sun-dried raisins made from different fragrance intensities grape varieties using a validated HS-SPME with GC-MS method. Food Chemistry. 2017;**228**:125-135

[112] Zhang H, Xie Y, Liu C, Chen S, Hu S, Xie Z, et al. Comprehensive comparative analysis of volatile compounds in citrus fruits of different species. Food Chemistry. 2017;**230**: 316-326

[113] Giannetti V, Mariani MB, Mannino P, Marini F. Volatile fraction analysis by HS-SPME/GC-MS and chemometric modeling for traceability of apples cultivated in the Northeast Italy. Food Control. 2017;**78**: 215-221

[114] Chen Q, Song J, Bi J, Meng X, Wu X. Characterization of volatile profile from ten different varieties of Chinese jujubes by HS-SPME/GC-MS coupled with E-nose. Foodservice Research International. 2018;**105**: 605-615

[115] Nijssen LM, Maarse H. Volatile compounds in black currant products: An additional factor in authenticity

control of fruit juices. Flavour and Fragrance Journal. 1986;**1**:143-148

[116] Low NH. Determination of fruit juice authenticity by capillary gas chromatography with flame ionization detection. Journal of AOAC International. 1996;**79**:724-737

[117] Del Castillo MLR, Caja MM, Herraiz M. Use of the enantiomeric composition for the assessment of the authenticity of fruit beverages. Journal of Agricultural and Food Chemistry. 2003;**51**:1284-1288

[118] Reinhard H, Sager F, Zoller O. Citrus juice classification by SPME-GC-MS and electronic nose measurements. LWT- Food Science and Technology. 2008;**41**:1906-1912

[119] Cuevas FJ, Pereira-Caro G, Moreno-Rojas JM, Munoz-Redondo JM, Ruiz-Moreno MJ. Assessment of premium organic orange juices authenticity using HPLC-HR-MS and HS-SPME-GC-MS combining data fusion and chemometrics. Food Control. 2017;**82**:203-211

[120] Armanino C, De Acutis R, Festa MR. Wheat lipids to discriminate species, varieties, geographical origins and crop years. Analytica Chimica Acta. 2002;**454**:315-326

[121] Pelillo M, Iafelice G, Marconi E, Caboni MF. Identification of plant sterols in hexaploid and tetraploid wheats using gas chromatography with mass spectrometry. Rapid Communications in Mass Spectrometry. 2003;**17**:2245-2252

[122] Tres A, Heenan SP, van Ruth S. Authentication of dried distilled grain with solubles (DDGS) by fatty acid and volatile profiling. LWT- Food Science and Technology. 2014;**59**:215-221

[123] Pastor K, Ačanski M, Vujić D, Bekavac G, Milovac S, Kravić S. Rapid method for small grain and corn flour authentication using GC/EI-MS and multivariate analysis. Food Analytical Methods. 2016a;**8**:443-450

[124] Pastor K, Acanski M, Vujic D, Kondic-Špika A. Binary simple sugar profiling in corn and small grain flour authentication using GC/EI-qMS approach. Chromatographia. 2016b;**79**: 1553-1559

[125] Pastor K, Acanski M, Vujic D, Jovanovic D, Wienkoop S. Authentication of cereal flours by multivariate analysis of GC-MS data. Chromatographia. 2016c;**79**:1387-1393

[126] Pastor K, Pezo L, Vujić D, Jovanović D, Ačanski M. Discriminating cereal and pseudocereal species using a binary system of GC-MS data—A pattern recognition approach. Journal of the Serbian Chemical Society. 2018;**83**: 317-329

[127] Psodorov Đ, Vujić D, Ačanski M, Pastor K, Razmovski R, Kravić S. The content of buckwheat flour in wheat bread. Acta Periodica Technologica. 2014;**45**:79-87

[128] Psodorov D, Ačanski M, Psodorov D, Vujić D, Pastor K. Determination of the content of buckwheat and wheat flours in bread using GC-MS and multivariate analysis. Journal of Food and Nutrition Research. 2015;**54**:179-183

[129] Zhang Z, Li T, Wang D, Zhang L, Chen G. Study on the volatile profile characteristics of oyster *Crassostrea gigas* during storage by a combination sampling method coupled with GC/MS. Food Chemistry. 2009;**115**:1150-1157

[130] Iglesias J, Medina I, Bianchi F, Careri M, Mangia A, Musci M. Study of the volatile compounds useful for the characterisation of fresh and frozen-thawed cultured gilthead sea bream fish by solid-phase microextraction gas

chromatography-mass spectrometry. Food Chemistry. 2009;**115**:1473-1478

[131] Zhang Z, Li G, Luo L, Chen G. Study on seafood volatile profile characteristics during storage and its potential use for freshness evaluation by headspace solid phase microextraction coupled with gas chromatography-mass spectrometry. Analytica Chimica Acta. 2010;**659**:151-158

[132] van Houcke J, Medina I, Linssen J, Luten J. Biochemical and volatile organic compound profile of European flat oyster (*Ostrea edulis*) and Pacific cupped oyster (*Crassostrea gigas*) cultivated in the Eastern Scheldt and Lake Grevelingen, the Netherlands. Food Control. 2016;**68**:200-207

[133] Nurjuliana M, Che Man YB, Mat Hashim D, Mohamed AKS. Rapid identification of pork for halal authentication using the electronic nose and gas chromatography mass spectrometer with headspace analyzer. Meat Science. 2011;**88**:638-644

[134] Gorska-Horczyczak E, Horczyczak M, Guzek D, Wojtasik-Kalinowska I, Wierzbicka A. Chromatographic fingerprints supported by artificial neural network for differentiation of fresh and frozen pork. Food Control. 2017;**73**:237-244

[135] Rui Alves M, Casal S, Oliveira MBPP, Ferreira MA. Contribution of FA profile obtained by high-resolution GC/chemometric techniques to the authenticity of green and roasted coffee varieties. JAOCS. 2003;**80**:511-517

[136] Risticevic S. HS-SPME-GC-TOFMS methodology for verification of geographical origin and authenticity attributes of coffee samples [thesis]. Ontario, Canada: Waterloo; 2008

[137] Lienert D, Anklam E, Panne U. Gas chromatography-mass spectral analysis of roots of *Echinacea* species and classification by multivariate data analysis. Phytochemical Analysis. 1998; **9**:88-98

[138] Lin J, Zhang P, Pan Z, Xu H, Luo Y, Wang X. Discrimination of oolong tea (*Camellia sinensis*) varieties based on feature extraction and selection from aromatic profiles analysed by HS-SPME/GC-MS. Food Chemistry. 2013; **141**:259-265

[139] Farag M, Rasheed DM, Kamal IM. Volatiles and primary metabolites profiling in two *Hibiscus sabdariffa* (roselle) cultivars via headspace SPME-GC-MS and chemometrics. Foodservice Research International. 2015;**78**:327-335

[140] Zhu Y, Shao C-Y, Lva H-P, Zhang Y, Dai W-D, Guo L, et al. Enantiomeric and quantitative analysis of volatile terpenoids indifferent teas (*Camellia sinensis*). Journal of Chromatography. A. 2017;**1490**:177-190

[141] Cui S, Wu J, Wang J, Wang X. Discrimination of American ginseng and Asian ginseng using electronic nose and gas chromatography mass spectrometry coupled with chemometrics. Journal of Ginseng Research. 2017;**41**:85-95